PERSONAL BRANDING

A COMPLETE SYSTEM TO BUILD AND MONETIZE A UNIQUE BRAND USING INFLUENCER STRATEGIES AND STEP-BY-STEP MARKETING TOOLS ON LINKEDIN, X, INSTAGRAM, YOUTUBE, FACEBOOK, AND TIKTOK

DAVE STONE

TOP TIER PRESS

Contents

INTRODUCTION 1

The Shift from Corporate Identity to Personal Identity in Business 1

Why Personal Branding Is Non-Negotiable in the Digital Age 2

What This Book Provides: A Complete System from Building to Monetizing 3

Reader Promise: Transform Visibility into Influence and Influence into Income 4

Unlock Your Social Authority Playbook 7

1. DISCOVERING YOUR UNIQUE BRAND IDENTITY 9

Defining Your Mission and Message 12

Crafting Your Story for Impact and Relatability 14

2. CRAFTING A MAGNETIC BRAND MESSAGE 19

Identifying Your Target Audience and Ideal Followers 22

Developing a Consistent Voice and Style 24

3. DESIGNING YOUR BRAND LOOK AND FEEL 29

Creating a Professional Online Presence 32

Brand Consistency Across Platforms 34

4. BUILDING AUTHORITY ON LINKEDIN 39

Publishing Content That Positions You as a Thought Leader 42

Leveraging Connections, Networking, and Lead Generation 45

5. BECOMING A VOICE ON X (TWITTER) 49

Threads and Viral Strategies 51

Engaging in Industry Conversations for Visibility 53

6. VISUAL STORYTELLING ON INSTAGRAM 59

Reels, Stories, and Carousels That Build Trust and Relatability 62

Converting Followers into Loyal Fans or Clients 65

7. LONG-FORM INFLUENCE ON YOUTUBE 69

Balancing Entertainment with Education 72

Monetization Through Ads, Sponsorships, and Products 74

8. COMMUNITY AND TRUST ON FACEBOOK 79
Live Sessions for Relationship Building 82
Using Facebook as a Funnel for Business Growth 84

9. TIKTOK FOR FAST-TRACK GROWTH 89
Creating Short-Form Content with High Virality Potential 92
Turning Trends into Authority-Driven Content 94

10. CONTENT STRATEGY ACROSS PLATFORMS 99
Scheduling, Consistency, and Automation 102
Analytics: How to Measure Growth and Impact 104

11. MONETIZING YOUR PERSONAL BRAND 109
Building Credibility to Attract Partnerships 112
Creating a Long-term Brand Monetization Roadmap 114

12. SCALING AND SUSTAINING YOUR BRAND 119
Delegating and Outsourcing Brand Management 121
Building a Brand That Thrives Beyond Trends 123

Your Words Could Inspire the Next Great Personal Brand 129

CONCLUSION 131
The Shift from Corporate Identity to Personal Identity in Business 131
Why Personal Branding Is Essential in the Digital Age 132
What This Book Provides: A Complete System from Building to Monetizing 133
Reader Promise: Transform Visibility into Influence and Influence into Income 134

Introduction

The rules of business have fundamentally changed. Gone are the days when hiding behind corporate logos and company brands was enough to build a successful career or business. Today, people buy from people, not faceless corporations. They follow individuals who share their values, solve their problems, and inspire their aspirations.

THE SHIFT FROM CORPORATE IDENTITY TO PERSONAL IDENTITY IN BUSINESS

Traditional business operated on institutional trust. Companies spent decades building brand recognition through advertising, public relations, and corporate messaging. Employees were expected to be interchangeable parts in the corporate machine, their individual personalities absorbed into company culture and brand guidelines.

This model has become obsolete in the digital age. Consumers now expect transparency, authenticity, and personal connection from the brands they support. They want to know who stands behind the products and services they purchase. They seek relationships, not just transactions.

Consider how business decisions are made today. When choosing between two consultants with similar credentials, clients almost always select the one whose personal brand resonates with their values and communication style. When investors evaluate startups, they are betting on the founders' personal brands as much as the business concept. When companies hire executives, they consider not just professional qualifications but also how that person's personal brand aligns with and enhances their organizational goals.

The most successful professionals understand that their personal brand is their competitive advantage. It is what differentiates them in crowded markets, attracts opportunities, and commands premium pricing. Your personal brand becomes your business moat, something competitors cannot replicate because it is uniquely yours.

WHY PERSONAL BRANDING IS NON-NEGOTIABLE IN THE DIGITAL AGE

Digital platforms have democratized influence and entrepreneurship in ways never seen before. Anyone with expertise, passion, and strategic thinking can build a global audience and generate significant income. However, this same accessibility has created intense competition for attention and trust.

Without a strong personal brand, you become invisible in the digital noise. Your expertise remains hidden, your potential clients cannot find you, and opportunities pass to those who have mastered the art of personal branding. The choice is not whether to build a personal brand, but whether to build one intentionally or allow others to define it for you.

Search engines and social media algorithms favor individuals who consistently share valuable content and engage authentically with their communities. They reward personal brands that demonstrate expertise, build relationships, and create genuine value for their audiences. These platforms have become meritocracies where the most helpful, authentic, and consistent personal brands rise to the top.

Modern consumers conduct extensive research before making purchasing decisions. They examine your online presence, read your content, watch your videos, and evaluate your credibility before engaging with your services or products. Your personal brand serves as your twenty-four-hour sales representative, building trust and demonstrating value even when you are not directly involved in the conversation.

The global nature of digital business means your personal brand can attract opportunities from anywhere in the world. Geographic limitations no longer constrain professional growth when your expertise and personality can reach global audiences through strategic personal branding.

WHAT THIS BOOK PROVIDES: A COMPLETE SYSTEM FROM BUILDING TO MONETIZING

This book provides a comprehensive, step-by-step system for building and monetizing a personal brand across all major digital platforms. You will learn not just what to do, but how to do it efficiently and effectively, with specific strategies tailored to each platform's unique characteristics and audience expectations.

The system covers every aspect of personal branding: discovering your unique value proposition, crafting compelling messages, designing professional visual elements, building authority across multiple platforms, creating engaging content consistently, and converting your influence into sustainable income streams.

Each chapter builds upon the previous one, creating a cohesive strategy that compounds your efforts rather than scattering them across disconnected tactics. You will understand how content created for one platform can be adapted and repurposed for others, how different platforms serve different functions in your overall brand strategy, and how to create synergies that amplify your impact.

The strategies presented here are based on proven methods used by successful personal brands across various industries. They are practical, implementable, and designed to generate real results rather than just theoretical understanding. Every concept is accompanied by specific action steps that allow immediate implementation.

READER PROMISE: TRANSFORM VISIBILITY INTO INFLUENCE AND INFLUENCE INTO INCOME

By the end of this book, you will have a complete roadmap for transforming from hidden expert to visible authority to profitable influencer. This transformation does not happen overnight, but it follows predictable patterns that you can accelerate through strategic action and consistent execution.

You will understand how to identify and articulate your unique value in ways that resonate with your target audience. You will know how to create content that builds trust, demonstrates expertise, and naturally leads to business opportunities. You will have specific strategies for each major social platform and understand how to integrate them into a cohesive personal branding system.

Most importantly, you will have a clear path for monetizing your personal brand through multiple income streams including consulting, speaking, digital products, sponsorships, and partnership opportunities. You will understand how to price your expertise, attract premium clients, and build a sustainable business around your personal brand.

The journey from anonymity to influence requires commitment and consistent action, but the rewards extend far beyond financial success. A strong personal brand provides professional security, creative fulfillment, and the satisfaction of helping others while building something uniquely yours.

Your expertise has value. Your perspective matters. Your audience is waiting for someone exactly like you to help them achieve their goals. This book provides the systematic approach to connect your unique value with the people who need it most, creating mutual benefit that forms the foundation of sustainable business success.

Unlock Your Social Authority Playbook

Ready to turn your online presence into lasting influence and income?

You've mastered the basics. Now it's time to go beyond. **The Social Authority Playbook** gives you clear systems and tools to scale your brand, earn more, and grow with confidence. Inside, you'll find proven tactics to automate your content, expand your reach, and build genuine trust that lasts.

What you'll discover inside::

- Build loyal digital communities
- Streamline your content systems
- Monetize your personal expertise
- Future proof your online brand

Access Your Playbook Now

Visit: **https://content.toptierpress.com/amplify-online-influence**

Or scan the QR code below to get instant access!

ONE

Discovering Your Unique Brand Identity

Building a powerful personal brand begins with deep self-understanding. Most professionals skip this foundational step and jump directly to content creation or platform building, which explains why their brands feel generic and fail to attract engaged audiences. Your unique combination of experiences, knowledge, and perspectives cannot be replicated by anyone else, but you must first identify and articulate what makes you distinctive.

Effective self-assessment goes beyond listing job titles or educational credentials. It involves examining the intersection of your natural talents, developed skills, genuine interests, and life experiences that have shaped your worldview. This intersection is where your most authentic and compelling personal brand emerges.

The goal isn't to fabricate a persona or present a perfect image, but to identify the genuine aspects of your background and personality that provide unique value to others. Authenticity resonates because people instinctively recognize and trust genuine expertise paired with authentic personality.

Identifying Your Core Strengths

Your core strengths include both natural talents and developed capabilities that consistently produce superior results. These aren't just skills you possess, but areas where you naturally excel and others frequently seek your input or assistance.

Consider the compliments you receive regularly from colleagues, friends, or clients. What do people consistently praise about your work or approach? What tasks do others find challenging that come naturally to you? These patterns reveal strengths that you might take for granted but represent significant value to others.

Examine your professional history for recurring themes and consistent successes. Look beyond job titles to identify the specific contributions that generated the best outcomes in each role. What processes did you improve? What problems did you solve? What results did you achieve that others struggled to replicate?

Don't overlook strengths that seem unrelated to your current professional focus. Communication skills developed through teaching, organizational abilities honed through event planning, or analytical thinking refined through personal projects can all contribute to your unique professional value proposition.

Uncovering Your Genuine Passions

Passions provide the emotional fuel that sustains long-term personal brand building. Without genuine interest in your chosen focus area, you'll struggle to create consistent content, engage authentically with your audience, and maintain the persistence required for brand building success.

Identify topics that naturally capture your attention and energy. What subjects do you research voluntarily? What conversations energize rather than drain you? What problems do you solve in your spare time because you find them intellectually stimulating rather than burdensome?

Consider the intersection between your passions and market demand. The most sustainable personal brands operate at the convergence of what you love doing, what you're naturally good at, and what others value enough to pay for. This intersection ensures both personal satisfaction and business viability.

Don't assume your passions must align perfectly with traditional career paths. Many successful personal brands are built around unique combinations of interests that create distinctive niches. A fitness enthusiast with financial expertise might build a brand around financial wellness, while a technology professional with artistic interests might focus on creative applications of emerging technologies.

Cataloging Your Expertise and Experience

Your expertise extends far beyond formal credentials to include practical experience, problem-solving capabilities, and insights gained through real-world application of your knowledge. This practical wisdom often provides more value to your audience than theoretical knowledge alone.

Document your professional journey with attention to specific challenges you've navigated, solutions you've developed, and lessons you've learned through both successes and failures. These experiences provide the raw material for compelling content that resonates with others facing similar challenges.

Include expertise gained outside traditional professional settings. Volunteer work, personal projects, hobby pursuits, and life experiences all contribute to your unique knowledge base. Parents develop project management and conflict resolution skills, travelers gain cultural understanding and adaptability, and hobbyists often possess deep technical knowledge in specialized areas.

Consider the problems you solve naturally for friends, family, or colleagues. What questions do people ask you repeatedly? What situations do others find stressful that you handle easily? These patterns indicate areas of expertise that you might not formally recognize but that provide significant value to others.

DEFINING YOUR MISSION AND MESSAGE

Your mission and message form the foundation of everything you communicate through your personal brand. Without clarity about your fundamental purpose and core message, your content will lack focus and your audience will struggle to understand how you can help them achieve their goals.

A compelling mission statement answers why you do what you do and whom you serve, while your core message articulates the unique value you provide and the transformation you help others achieve. Together, these elements guide every decision about content creation, audience engagement, and business development.

The most powerful missions and messages emerge from the intersection of your personal experiences, professional expertise, and genuine desire to help others overcome challenges or achieve aspirations. This intersection ensures authenticity while creating clear value propositions that attract engaged audiences.

Crafting Your Mission Statement

Your mission statement should be specific enough to guide decision-making while broad enough to encompass your range of expertise and interests. It should clearly identify whom you serve and what transformation or outcome you help them achieve.

Effective mission statements focus on the impact you create rather than the services you provide. "I help overwhelmed entrepreneurs build systematic business operations" is more compelling than "I provide business consulting services" because it speaks directly to the audience's desired outcome.

Consider the deeper motivation behind your professional work. What injustice do you want to correct? What inefficiency do you want to eliminate? What possibility do you want to help others realize? This deeper purpose provides the emotional foundation that sustains long-term brand-building efforts.

Test your mission statement against your natural conversations about your work. When you're genuinely excited about helping someone, what do you say about what you do and why it matters? This natural language often provides better foundation material than formal business jargon.

Developing Your Core Message

Your core message distills your unique value proposition into clear, memorable language that immediately communicates why someone should pay attention to you rather than the many other experts in your space.

Strong core messages include specific outcomes, target audiences, and unique approaches or perspectives that differentiate you from others addressing similar problems. "I help technology leaders build inclusive teams through data-driven hiring practices" is more compelling than "I help companies with diversity and inclusion."

Your core message should be validated by your actual experience and results. If you claim to help businesses double their revenue, you should have concrete examples of achieving this outcome for clients. If you position yourself as an expert in work-life balance, your own life should demonstrate the principles you advocate.

Consider the language your ideal clients use to describe their challenges and aspirations. Your core message should speak their language and address their specific pain points using terminology they recognize and respond to emotionally.

Aligning Mission and Message with Market Needs

The most successful personal brands operate at the intersection of personal passion, unique expertise, and genuine market demand. All three elements must be present for sustainable brand-building success.

Research your target market to understand their most pressing challenges, unfulfilled needs, and desired outcomes. What problems keep them awake at night? What goals do they struggle to achieve? What solutions have they tried that failed to deliver promised results?

Validate demand for your unique approach through direct conversation with potential clients or audience members. Do they recognize the problems you solve? Do they value the outcomes you help achieve? Are they willing to invest time, attention, or money in solutions to these challenges?

Consider how your mission and message fit within the broader conversation in your industry or niche. What perspectives are missing? What approaches aren't being tried? What audiences aren't being served effectively by existing experts?

CRAFTING YOUR STORY FOR IMPACT AND RELATABILITY

Your personal story transforms abstract expertise into relatable human experience that creates emotional connections with your audience. People don't just buy products or services, they buy from individuals they trust, respect, and relate to on a personal level.

Effective brand storytelling doesn't require dramatic transformation narratives or extraordinary circumstances. The most compelling stories often emerge from ordinary challenges that your audience faces, showing how you navigated these difficulties and what you learned in the process.

Your story should highlight the journey from where you started to where you are now, with particular attention to the struggles, lessons, and insights that shaped your current expertise and perspective. This journey provides context for your knowledge while demonstrating empathy for others facing similar challenges.

Identifying Your Transformation Narrative

Every compelling personal brand story includes elements of transformation. How you moved from one state to another through effort, learning, and growth. This transformation doesn't need to be dramatic, but it should be genuine and relevant to your audience's aspirations.

Consider significant challenges you've overcome in your personal or professional life. What obstacles seemed insurmountable at the time? What fears did you face? What mistakes did you make? What breakthroughs changed your trajectory? These elements provide raw material for authentic storytelling.

Focus on transformations that relate to the problems you now help others solve. If you help businesses improve their operations, share how you learned to systematize your own work or organizational processes. If you teach productivity strategies, describe your journey from chaos to organized effectiveness.

The most powerful transformation narratives include details that make the story vivid and memorable. Rather than saying "I struggled with time management," describe the specific consequences of poor organization and the moment you realized something had to change.

Structuring Your Story for Maximum Impact

Effective brand stories follow predictable structures that create emotional engagement and memorable impact. The classic transformation arc moves from initial challenge through struggle and discovery to breakthrough and current state.

Begin with relatable circumstances that your audience recognizes—the problems, frustrations, or aspirations that many people in your target market experience. This creates an immediate connection and demonstrates your understanding of their situation.

Describe the journey of attempting to solve these problems, including failures, partial successes, and gradual learning that led to your current expertise. This journey phase shows persistence, learning ability, and the development of hard-won wisdom that provides value to others.

Conclude with the transformation you achieved and how this experience now enables you to help others navigate similar challenges more effectively. This conclusion positions you as a guide who has walked the path rather than a theorist who studied the map.

Making Your Story Relatable and Actionable

The best brand stories provide inspiration while offering practical insights that audience members can apply to their situations. Avoid stories that seem so unique or extraordinary that others cannot relate to your experience or imagine achieving similar outcomes.

Include specific strategies, mindset shifts, or practical actions that contributed to your transformation. What books did you read? What habits did you develop? What people influenced your thinking? What mistakes taught you valuable lessons? These details provide actionable guidance within your narrative.

Address common objections or limiting beliefs that might prevent others from taking action. If someone might think, "That worked for you, but it won't work for me," explain how your approach can be adapted to different circumstances or personality types.

Connect your story to universal themes and emotions that transcend specific industries or circumstances. Stories about overcoming fear, developing confidence, learning from failure, or discovering purpose resonate across diverse audiences while maintaining professional relevance.

Action Steps for Chapter 1:

Action Item	Task Description	Expected Outcome
Strengths and Expertise Inventory	Create a comprehensive list of your professional skills, natural talents, and unique experiences. Include feedback from others and recurring themes across your career journey.	Clear understanding of your unique value and competitive advantages
Mission Statement Development	Write a one-sentence mission statement that identifies who you serve and what transformation you help them achieve. Test it with trusted colleagues for clarity and impact.	Focused foundation that guides all brand decisions and communications
Core Message Crafting	Develop a 30-second elevator pitch that communicates your unique value proposition. Practice delivering it naturally in various professional situations.	Memorable, clear communication of what makes you distinctive and valuable
Transformation Story Documentation	Write your personal transformation story following the challenge-journey-breakthrough structure. Focus on experiences relevant to your target audience's aspirations.	Compelling narrative that builds trust and demonstrates your expertise through real experience
Market Validation Research	Conduct informal interviews with 5–10 people in your target market to validate that your mission and message address real needs and resonate with your intended audience.	Confirmation that your brand identity aligns with genuine market demand

TWO

Crafting a Magnetic Brand Message

Your elevator pitch serves as the cornerstone of all your brand communications. It's the distilled essence of who you are, what you do, and why it matters, delivered in a way that captures attention and invites further conversation. Unlike generic networking introductions that blend into forgettable noise, a magnetic elevator pitch creates immediate interest and positions you as someone worth knowing.

The most effective elevator pitches focus on the transformation you create rather than the services you provide. People don't buy consulting; they buy better business results. They don't hire coaches; they hire someone who helps them achieve their goals faster and more efficiently. Your pitch should immediately connect your expertise to outcomes your audience cares about.

A well-crafted elevator pitch adapts to different contexts while maintaining consistent core messaging. The version you use at industry conferences might emphasize different aspects than what you share at community networking events, but both should convey the same fundamental value proposition and leave listeners with a clear understanding of how you might help them.

The Problem-Solution-Outcome Formula

Structure your elevator pitch around the proven problem-solution-outcome formula that immediately establishes relevance and value. Begin by identifying a problem your audience faces, present your unique solution approach, and conclude with the specific outcome you help achieve.

"Most small business owners struggle with inconsistent marketing results because they're trying random tactics instead of following proven systems. I help them build marketing funnels that generate predictable leads and convert prospects into paying clients, typically increasing their revenue by 30-50% within six months."

This structure works because it follows natural human thinking patterns. People are drawn to those who understand their challenges, interested in solutions that make sense, and motivated by outcomes they want to achieve. The formula creates engagement at each step while building toward a compelling conclusion.

Avoid starting with your job title or credentials, which provide context but don't create emotional connection. "I'm a marketing consultant with an MBA and ten years of experience" tells people what you are, not what you do for them. Lead with the problem you solve, and credentials become supporting evidence rather than opening statements.

Quantifying Your Impact

Specific numbers and measurable outcomes transform vague promises into credible claims that capture attention and build confidence. Instead of saying you help people "improve their performance," specify that you "help executives increase their team's productivity by 25% while reducing overtime costs."

Use metrics that matter to your target audience. Financial professionals care about ROI and profit margins. Operations managers focus on efficiency and cost reduction. Marketing leaders want engagement rates and conversion percentages. HR executives value retention rates and employee satisfaction scores.

When you don't have specific client results to reference, use industry benchmarks or research data that supports your approach. "Studies show that companies with systematic onboarding processes see 60% better new hire retention," which provides a credible foundation for your onboarding expertise even without specific client case studies.

Be prepared to support your claims with brief examples or explanations when asked. Your elevator pitch creates interest, but follow-up questions require more detailed validation of your stated outcomes and approach.

Adapting Your Pitch for Different Audiences

Your core message remains consistent, but the emphasis and language should adapt to different audience contexts and interests. The problems that matter most to startup founders differ from those facing corporate executives, even when both might benefit from your expertise.

Research your audience before networking events or professional gatherings. What challenges does this industry face? What language do they use to describe their problems? What outcomes do they value most highly? This preparation allows you to customize your pitch for maximum relevance and impact.

Practice variations of your pitch that emphasize different aspects of your value. One version might focus on cost savings for budget-conscious audiences, while another emphasizes growth opportunities for expansion-minded leaders. The core transformation remains the same, but the framing adapts to different motivations.

Listen actively to how others introduce themselves and describe their challenges. This real-time market research helps you adjust your pitch to address the specific concerns and interests you hear expressed in that environment.

IDENTIFYING YOUR TARGET AUDIENCE AND IDEAL FOLLOWERS

Building a powerful personal brand requires absolute clarity about who you're trying to reach and influence. Many professionals try to appeal to everyone, but this results in messages that resonate with no one. Specific audiences with clearly defined needs, challenges, and aspirations form the foundation of the most successful personal brands.

Your target audience isn't just anyone who might potentially benefit from your expertise. It's the specific group of people you can help most effectively and who are most likely to engage with your content, share your message, and eventually become clients or customers.

Effective audience identification goes beyond demographic categories to include psychographic factors like values, beliefs, goals, frustrations, and communication preferences. Two people with identical job titles might have completely different attitudes toward professional development, risk tolerance, or decision-making processes.

Creating Detailed Audience Personas

Develop comprehensive profiles of your ideal audience members that go far beyond basic demographics. Include their professional responsibilities, personal goals, daily challenges, information consumption habits, and preferred communication styles.

Consider where your ideal followers spend their time online and offline. What publications do they read? What conferences do they attend? What social media platforms do they use most actively? What podcasts do they listen to during their commute? Understanding their media consumption patterns helps you position your content where they're most likely to discover it.

Document the language your audience uses to describe their challenges and aspirations. Do they talk about "work-life balance" or "professional fulfillment"? Do they focus on "increasing revenue" or "building sustainable businesses"? Using their language in your messaging creates immediate resonance and demonstrates understanding.

Include emotional and psychological factors in your audience profiles. What keeps them awake at night? What makes them feel most confident and capable? What fears hold them back from taking action? What aspirations motivate their professional decisions? These insights inform content that creates genuine connection.

Understanding Audience Pain Points and Desires

Your audience's pain points represent opportunities to provide immediate value through your content and expertise. The more accurately you can identify and articulate their specific challenges, the more your brand will resonate as the solution they've been seeking.

Distinguish between surface-level complaints and underlying root causes. Someone might complain about "not having enough time," but the real issue could be poor prioritization skills, lack of delegation capabilities, or inefficient systems and processes. Address the root causes, not just the symptoms.

Research the solutions your audience has already tried and why those approaches failed or fell short. This insight helps you position your approach as superior to existing alternatives while acknowledging their previous efforts and frustrations.

Identify aspirational outcomes that motivate your audience's decisions. What does success look like to them? What would their ideal professional life include? What recognition or achievements do they dream about? Understanding their positive motivations helps you frame your value proposition in inspiring rather than just problem-solving terms.

Analyzing Where Your Audience Congregates

Map the online and offline spaces where your ideal audience members gather to learn, network, and discuss industry topics. These locations become prime opportunities for building visibility and establishing your expertise within relevant communities.

Professional platforms like LinkedIn groups, industry forums, specialized communities, and association websites provide opportunities to engage directly with your target audience. Observe the conversations, questions, and concerns that generate the most engagement to understand what topics matter most.

Conference attendee lists, speaker lineups, and event agendas reveal where your audience invests their professional development time and budgets. These events provide networking opportunities while also indicating topics and themes that resonate with your target market.

Social media analytics from competitors and industry leaders show which platforms generate the most engagement from your target demographic. If your audience is most active on LinkedIn during business hours but engages with video content on YouTube during evenings and weekends, this information guides your content distribution strategy.

DEVELOPING A CONSISTENT VOICE AND STYLE

Your brand voice encompasses the personality, tone, and communication style that makes your content immediately recognizable and distinctly yours. Consistency in voice builds trust, creates expectation, and helps your audience develop a relationship with your brand that extends beyond individual pieces of content.

Developing an authentic brand voice requires understanding both your natural communication style and your audience's preferences and expectations. The goal isn't to create an artificial persona but to amplify the aspects of your authentic personality that resonate most effectively with the people you want to reach and influence.

Your voice should feel sustainable over time and across different content formats. If maintaining your chosen voice requires constant effort or feels unnatural, it will become inconsistent and ultimately ineffective. The best brand voices enhance rather than mask your authentic personality.

Defining Your Voice Characteristics

Identify three-to-five specific adjectives that capture the essence of how you want to communicate with your audience. These might include professional yet approachable, confident but humble, practical and results-focused, or innovative and forward-thinking. These characteristics guide every communication decision.

Consider the relationship dynamic you want to create with your audience. Do you want to be seen as the authoritative expert, the helpful peer, the challenging coach, or the encouraging mentor? This positioning influences your tone, language choices, and the level of vulnerability or authority you express.

Analyze your most successful communications, emails, presentations, or conversations that generated positive responses and strengthened relationships. What voice characteristics made those interactions effective? How can you consistently replicate those elements across all your brand communications?

Document specific language preferences and communication patterns that align with your brand voice. Do you use formal or conversational language? Do you prefer direct statements or questions that encourage engagement? Do you include humor, personal anecdotes, or stick to professional insights? These preferences create consistency across all your content.

Adapting Voice Across Platforms

While your core voice remains consistent, the expression of that voice adapts to different platform cultures and audience expectations. LinkedIn content might emphasize the professional and strategic aspects of your voice, while Instagram could highlight the personal and approachable elements.

Understand each platform's communication norms and user expectations. Twitter favors concise, often witty observations. LinkedIn rewards thoughtful, professional insights. YouTube allows for more casual, conversational communication. Instagram combines visual storytelling with authentic personality expression.

Maintain voice consistency through underlying values and perspectives rather than identical language and tone. Your core beliefs, expertise areas, and communication goals remain constant while the delivery method adapts to platform-specific requirements.

Test different voice expressions on various platforms to identify what generates the best engagement and feels most authentic to you. Some professionals discover they're naturally more effective with video content because their personality translates better through visual and audio cues than written text alone.

Creating Voice Guidelines and Standards

Document your voice characteristics, preferred language, and communication standards in a reference guide that ensures consistency across all your brand communications. This becomes especially important when you begin delegating content creation or collaborating with others on your brand development.

Include specific examples of how your voice handles different types of content—educational posts, promotional messages, responses to criticism, celebratory announcements, and personal shares. These examples provide templates for maintaining consistency across diverse communication scenarios.

Identify voice elements that conflict with your desired brand positioning. If you want to be seen as approachable, avoid overly technical jargon. If you're building authority, eliminate uncertain language like "I think" or "maybe." If you emphasize authenticity, avoid corporate speak and buzzwords.

Regularly review your content to ensure voice consistency and authenticity. Ask trusted colleagues or clients for feedback about how your brand voice comes across and whether it aligns with their experience of your personality and expertise in direct interactions.

Action Steps for Chapter 2:

Action Item	Task Description	Expected Outcome
Elevator Pitch Development	Create three versions of your elevator pitch using the problem-solution-outcome formula: a 30-second version, 60-second version, and 2-minute detailed version. Practice delivering each naturally in different scenarios.	Polished, compelling introduction that generates interest and positions you as a valuable expert
Audience Persona Creation	Develop detailed profiles of your top 3 ideal audience segments including demographics, psychographics, pain points, goals, and preferred communication channels. Interview actual people from each segment to validate your assumptions.	Deep understanding of who you're trying to reach and what messages will resonate most effectively
Voice Characteristics Definition	Identify 5 specific adjectives that describe your desired brand voice and create examples of how this voice handles different types of content (educational, promotional, personal, professional).	Clear, consistent communication style that feels authentic while appealing to your target audience

Action Item	Task Description	Expected Outcome
Message Testing Program	Test your elevator pitch and key messages with 10 people from your target audience. Gather feedback on clarity, relevance, and memorability to refine your core communications.	Validated messaging that resonates with real people and generates the responses you want
Platform Voice Adaptation	Adapt your core brand voice for each platform you plan to use, considering platform culture and audience expectations while maintaining consistency in values and perspective.	Platform-specific communication approaches that feel native while maintaining brand consistency

THREE

Designing Your Brand Look and Feel

Visual elements serve as the silent ambassadors of your personal brand, communicating professionalism, values, and personality before you speak a single word. In our image-saturated digital environment, strong visual branding isn't optional. It's essential for creating memorable first impressions and building recognition across platforms and interactions.

Your visual brand encompasses far more than just a logo. It includes your color palette, typography choices, photography style, graphic design elements, and the overall aesthetic that ties all your materials together. These elements work collectively to create a cohesive visual language that supports and reinforces your verbal messaging.

The most effective personal brand visuals strike a balance between professional credibility and authentic personality. Your visual choices should reflect both your expertise and your human warmth, creating an approachable yet authoritative presence that encourages connection while building trust.

Color Psychology and Palette Selection

Colors trigger emotional responses and psychological associations that operate below conscious awareness but significantly influence how others perceive your brand. Understanding color psychology allows you to make strategic choices that support your desired brand positioning and audience response.

Blue conveys trust, stability, and professionalism, making it popular among financial advisors, consultants, and corporate professionals. Red suggests energy, passion, and urgency, working well for motivational speakers, fitness experts, and sales professionals. Green implies growth, harmony, and prosperity, appealing to environmental consultants, wellness coaches, and financial planners.

However, avoid choosing colors based solely on industry conventions. Your color palette should align with your authentic personality and differentiate you from competitors who default to expected choices. A financial advisor who uses warm oranges and teals can stand out while still maintaining professionalism through thoughtful application.

Develop a primary palette of 2-3 colors that work harmoniously together and can be applied consistently across all your brand materials. Include neutral colors (whites, grays, or beiges) that provide breathing space and ensure your primary colors have maximum impact when used strategically.

Typography and Font Choices

Typography choices communicate personality and professionalism as powerfully as color selections. Fonts carry emotional weight and cultural associations that influence how others perceive your credibility and expertise before they read your content.

Serif fonts (like Times New Roman or Georgia) suggest tradition, reliability, and academic credibility, making them appropriate for lawyers, professors, or consultants who want to emphasize their scholarly expertise. Sans-serif fonts (like Arial or Helvetica) appear modern, clean, and approachable, working well for technology professionals, designers, or coaches who want to seem current and accessible.

Script fonts can add personality and warmth but should be used sparingly and only when legibility isn't compromised. They work best for creative professionals like photographers, designers, or lifestyle coaches who want to emphasize their artistic sensibility.

Establish a typography hierarchy that includes fonts for headlines, body text, and accent elements. Consistency in font choices across all your materials creates visual cohesion while ensuring your content remains readable across different platforms and devices.

Essential Visual Branding Elements:

- **Professional Headshots**: Invest in high-quality photography that captures your personality while maintaining professional standards. Your headshot should be consistent across all platforms and updated regularly to reflect your current appearance and brand evolution.
- **Logo Design**: Create a simple, memorable mark that works effectively at various sizes and across different backgrounds. Your logo should be scalable from business card size to billboard dimensions while maintaining clarity and impact.
- **Brand Pattern Library**: Develop consistent patterns, shapes, or design elements that create visual recognition and tie your materials together. These might include specific border styles, background textures, or graphic elements that appear across your content.
- **Image Filter/Style**: Establish consistent photo editing approaches that create visual cohesion across your content. This might include specific color adjustments, contrast levels, or filter applications that make your images immediately recognizable.

- **Template Systems**: Design templates for common content types (social media posts, presentations, documents) that ensure consistency while allowing efficient content creation. Templates save time while maintaining professional visual standards.
- **Brand Guidelines Document**: Create a reference guide that documents all your visual choices, including color codes, font specifications, logo usage rules, and application examples. This ensures consistency as your brand grows and evolves.

CREATING A PROFESSIONAL ONLINE PRESENCE

Your online presence serves as your digital headquarters, often providing the first comprehensive impression potential clients, employers, or collaborators have of your expertise and professionalism. Every element of your digital presence should work together to build credibility, demonstrate value, and encourage the next step in your professional relationship.

Creating a professional online presence extends beyond having profiles on social platforms. It includes your website, email signature, online portfolios, professional directories, and any other digital touchpoints where people might encounter your brand.

The goal isn't perfection but consistency and professionalism that accurately represents your capabilities while making it easy for others to understand how you might help them achieve their goals.

Website Development and Optimization

Your professional website serves as the central hub of your online presence, providing comprehensive information about your expertise while giving you complete control over the user experience and messaging.

Effective personal brand websites include clear value propositions, detailed service descriptions, client testimonials, comprehensive about pages, and obvious contact information. Visitors should understand within seconds what you do, whom you serve, and how to engage your services.

Include a professional bio that tells your story while emphasizing the experiences and credentials most relevant to your target audience. Your bio should balance professional accomplishments with personality elements that help visitors connect with you as a person.

Showcase your expertise through case studies, client results, media mentions, speaking engagements, and content samples. Social proof builds credibility while demonstrating the real-world impact of your work.

Optimize your website for search engines by including relevant keywords in your content, creating valuable resources that attract inbound links, and maintaining technically sound site performance. Many potential clients will discover you through search engines rather than social media or referrals.

Social Media Profile Optimization

Each social media platform requires platform-specific optimization that considers the unique culture and expectations of its user base while maintaining consistency with your overall brand message.

LinkedIn profiles should emphasize professional achievements, specific skills, and clear value propositions that help connection requests and business inquiries. Include keywords that your ideal clients might use when searching for your type of expertise.

Instagram profiles need compelling visuals, clear bio descriptions, and strategic use of highlights to showcase different aspects of your expertise. The visual nature of the platform requires more attention to aesthetic consistency and personality expression.

Twitter profiles should capture your expertise and personality in very limited space while demonstrating your engagement with industry conversations. Your tweet history becomes part of your profile, so maintain consistency with your desired brand positioning.

Ensure all your social profiles include professional headshots, consistent bio information, and clear contact information that makes it easy for interested parties to connect with you outside the platform.

Content Portfolio Development

Develop a comprehensive portfolio of your best work that demonstrates the range and quality of your expertise. This portfolio should be easily accessible and regularly updated with your most impressive recent accomplishments.

Include various content types that showcase different aspects of your capabilities: written articles that demonstrate thought leadership, video content that shows your communication skills, project results that prove your effectiveness, and testimonials that validate your client relationships.

Organize your portfolio strategically to guide visitors toward the type of engagement you most want to encourage. If you're seeking speaking opportunities, emphasize video content and previous speaking engagements. If you're building consulting business, highlight case studies and client results.

Make your portfolio easily shareable and accessible across all your professional platforms. Potential clients should be able to quickly access examples of your work regardless of where they first encounter your brand.

BRAND CONSISTENCY ACROSS PLATFORMS

Consistency across all platforms and touchpoints creates professional credibility while building recognition that compounds over time. When your visual elements, messaging, and personality remain consistent, people begin to recognize your content and associate it with quality and reliability.

However, consistency doesn't mean identical. Each platform has unique requirements and user expectations that require thoughtful adaptation while maintaining core brand elements that tie everything together.

The key is identifying which elements remain constant (your core colors, fonts, logo, and key messages) and which elements adapt to platform-specific requirements (image dimensions, content formats, and communication tone).

Platform-Specific Adaptation Strategies

LinkedIn requires a more formal, professional presentation that emphasizes business value and industry expertise. Your content should be comprehensive, well-researched, and focused on professional development or industry insights.

Instagram needs visually appealing content that showcases personality alongside expertise. The platform culture encourages more personal sharing and behind-the-scenes content that humanizes your professional brand.

Twitter favors concise, timely commentary on industry trends and current events. Your content should demonstrate thought leadership while engaging in broader conversations relevant to your expertise.

YouTube allows for comprehensive education and personality expression through longer-form content. Your videos should demonstrate teaching ability while showcasing both expertise and communication skills.

Facebook serves various functions depending on whether you use personal profiles, business pages, or groups. The platform works well for community building and more casual professional networking.

Visual Consistency Systems

Develop templates and style guides that ensure visual consistency while accommodating platform-specific requirements. Your Instagram posts might use different dimensions than your LinkedIn articles, but they should share color palettes, fonts, and design elements that create immediate recognition.

Create branded templates for common content types: quote graphics, educational carousels, video thumbnails, and presentation slides. These templates save time while ensuring professional appearance and brand consistency.

Maintain photo editing consistency through preset filters or editing approaches that create a cohesive visual style across all your imagery. Your photos should look like they belong to the same brand regardless of where they appear.

Use consistent graphic elements, color schemes, and typography across all platforms while adapting layouts and formats to meet platform-specific requirements and user expectations.

Message Consistency Framework

Your core value proposition, key messages, and fundamental positioning should remain consistent across all platforms while the presentation and emphasis adapt to different audience segments and platform cultures.

Develop a master message document that includes your elevator pitch, key talking points, value propositions, and core content themes. This document ensures consistency while providing flexibility for platform-specific adaptation.

Create content calendars that coordinate messaging across platforms, ensuring your audience receives consistent value regardless of where they follow your content. Avoid contradictory messages or positioning that might confuse your audience.

Regularly audit your brand presence across all platforms to identify inconsistencies and opportunities for improvement. Your brand should feel cohesive and professional regardless of where people encounter it.

Action Steps for Chapter 3:

Action Item	Task Description	Expected Outcome
Visual Brand Elements Creation	Develop your color palette (2–3 primary colors plus neutrals), select brand fonts, and create or commission a professional logo. Document all specifications in a brand guidelines reference.	Complete visual identity system that can be applied consistently across all brand materials
Professional Photography Session	Invest in professional headshots and brand photography that captures your personality while maintaining professional standards. Get multiple options for different contexts and platforms.	High-quality visual assets that enhance credibility and create strong first impressions
Website Development	Build or update your professional website with clear value propositions, comprehensive about page, service descriptions, testimonials, and contact information. Optimize for search engines and mobile devices.	Professional online headquarters that effectively represents your expertise and converts visitors to clients
Social Media Optimization	Update all your social media profiles with consistent branding elements, professional photos, optimized bio descriptions, and relevant contact information. Ensure platform-specific optimization while maintaining brand consistency.	Cohesive online presence that builds recognition and makes it easy for prospects to connect with you
Brand Guidelines Documentation	Create a comprehensive brand guide that documents all visual elements, messaging frameworks, and application examples. Include specific instructions for maintaining consistency across different contexts.	Reference system that ensures brand consistency as you create content and expand your presence

FOUR

Building Authority on LinkedIn

LinkedIn serves as the world's largest professional networking platform and represents the most important digital real estate for building business credibility and attracting high-value opportunities. Your LinkedIn profile functions as your professional headquarters, often providing the first comprehensive impression potential clients, employers, or collaborators have of your expertise and accomplishments.

Optimizing your LinkedIn profile requires strategic thinking beyond simply listing your job history. Every section should work together to tell a compelling story about your unique value while making it easy for your ideal audience to understand how you can help them achieve their goals.

The most effective LinkedIn profiles balance professional achievement with authentic personality, creating connections that feel both credible and approachable. Your profile should demonstrate expertise while showing the human side of your professional brand.

Headline Optimization for Maximum Impact

Your LinkedIn headline appears in search results, connection requests, and comment threads, making it one of your most valuable branding real estate opportunities. Rather than defaulting to your job title, craft a headline that communicates your value proposition and target audience.

"Marketing Director at TechCorp" tells people what you are but not what you do for others. "I help B2B SaaS companies increase qualified leads by 40% through strategic content marketing" immediately communicates value and attracts the right audience.

Include relevant keywords that your ideal connections might use when searching for expertise like yours. If you help companies with digital transformation, include terms like "digital strategy," "business transformation," or "technology implementation" that prospects might search for.

Keep your headline current and relevant to your goals. If you're transitioning careers, pivoting industries, or launching new services, your headline should reflect your desired positioning rather than just your current role.

About Section Storytelling

Your About section provides space for comprehensive storytelling that builds connection while demonstrating expertise. This section should balance professional accomplishments with personality elements that help visitors understand both your capabilities and your character.

Start with a compelling hook that immediately captures attention and speaks directly to your target audience's needs or aspirations. "After watching too many great ideas fail due to poor execution, I became obsessed with helping entrepreneurs build systematic business operations" creates immediate interest and positions expertise.

Include specific results and accomplishments that demonstrate your effectiveness. Quantify your impact wherever possible: "increased client retention rates by 35%," "generated $2.3M in new revenue," or "reduced operational costs by 22%." Numbers provide credible evidence of your capabilities.

Share your professional philosophy and approach in language that resonates with your target audience. What principles guide your work? What methods do you use that differ from others in your field? This insight helps prospects understand your unique approach and values alignment.

Conclude with a clear call-to-action that invites connection and engagement. "Let's connect if you're interested in discussing how systematic operations can accelerate your business growth" provides specific direction for interested visitors.

Experience Section Strategic Positioning

Present your work experience in ways that highlight transferable skills and relevant accomplishments rather than just listing job duties. Each role should demonstrate progression, learning, and results that support your current professional positioning.

Focus on achievements and outcomes rather than responsibilities and tasks. Instead of "managed marketing campaigns," write "developed content marketing strategy that increased lead generation by 65% while reducing cost-per-lead by 30%."

Include volunteer work, board positions, and significant projects that demonstrate relevant expertise and community involvement. These experiences often provide differentiation and show commitment to your professional field.

Use consistent formatting and professional language throughout your experience descriptions. Maintain active voice and quantify results wherever possible to create compelling evidence of your effectiveness.

Strategic LinkedIn Profile Optimization Tactics:

1. **Skills Section Curation**: Select and prioritize the 10-15 skills most relevant to your target audience and professional goals. Regularly update this section to reflect your current focus areas and encourage endorsements from colleagues and clients who can validate your expertise.
2. **Recommendations Strategy**: Proactively request recommendations from clients, colleagues, and supervisors who can speak specifically to your results and character. Offer to write recommendations for others first, which often generates reciprocal recommendations that enhance your credibility.
3. **Featured Section Utilization**: Use the Featured section to showcase your best work including articles, videos, presentations, or external content that demonstrates your expertise. This section provides visual interest while highlighting your most impressive professional accomplishments.
4. **Contact Information Optimization**: Make it easy for interested prospects to reach you by including multiple contact methods and links to your other professional platforms. Consider using a scheduling link if you regularly take discovery calls or consultations.

PUBLISHING CONTENT THAT POSITIONS YOU AS A THOUGHT LEADER

LinkedIn's publishing platform and native content features provide powerful opportunities to demonstrate expertise, share insights, and build authority within your professional community. Consistent, valuable content creation establishes you as a go-to resource while attracting the attention of potential clients, collaborators, and opportunities.

Thought leadership content differs from promotional material by focusing on industry insights, professional development, and valuable perspectives that help others succeed. The best LinkedIn content provides immediate value while subtly demonstrating the author's expertise and unique approach.

Content that performs well on LinkedIn balances professional value with authentic personality, creating posts that feel both informative and relatable. Your content should demonstrate not just what you know, but how you think and why others should care about your perspective.

Content Types That Build Authority

Industry analysis and trend commentary position you as someone who understands the broader context of your field and can help others navigate change. Share your perspective on industry developments, regulatory changes, market shifts, or emerging opportunities that affect your audience.

Case studies and client success stories (with appropriate permissions) demonstrate your ability to generate real results for others. These stories provide social proof while showing your methodology and approach in action.

Behind-the-scenes insights about your work process, decision-making approaches, or business operations humanize your expertise while providing valuable learning for others facing similar challenges.

Professional development content including lessons learned, skill-building advice, and career guidance appeals to LinkedIn's professional audience while demonstrating your experience and wisdom.

Content Structure for Maximum Engagement

Start your posts with compelling hooks that immediately capture attention and promise value. Questions, surprising statistics, bold statements, or personal anecdotes can draw readers into your content.

Use short paragraphs and line breaks to create scannable content that works well on mobile devices. LinkedIn users often browse during brief moments throughout their workday, so make your content easy to consume quickly.

Include specific examples, data, or stories that support your main points. Abstract advice becomes more valuable when illustrated with concrete applications that readers can relate to their own situations.

End posts with questions or calls-to-action that encourage engagement. The LinkedIn algorithm favors content that generates comments and discussion, so explicitly invite reader interaction.

Content Calendar and Consistency Strategy

Develop a content calendar that balances different types of posts while maintaining regular publishing frequency. Consistency builds audience expectations and algorithmic favor, but quality should never be sacrificed for quantity.

Plan content themes around your expertise areas, industry events, seasonal topics, and current news that affects your professional community. This strategic approach ensures relevance while demonstrating your thought leadership across various aspects of your field.

Repurpose your best content across different formats—turn a successful post into a longer article, create an infographic from popular insights, or develop video content around your most engaging written posts.

Monitor your content performance to identify topics, formats, and posting times that generate the best engagement with your specific audience. Use these insights to optimize future content strategies.

LEVERAGING CONNECTIONS, NETWORKING, AND LEAD GENERATION

LinkedIn's networking capabilities extend far beyond collecting connections to building meaningful professional relationships that generate business opportunities, collaborations, and career advancement. Strategic networking on LinkedIn requires systematic approaches that provide value to others while building your own professional network and opportunities.

The platform's advanced search capabilities, industry groups, and messaging features provide powerful tools for identifying, connecting with, and building relationships with your ideal professional contacts. However, success requires authentic relationship building rather than transactional networking approaches.

Effective LinkedIn networking focuses on providing value, building genuine relationships, and establishing yourself as a valuable connection worthy of referrals, collaborations, and business opportunities.

Strategic Connection Building

Identify specific types of professionals who could benefit from your expertise or provide valuable opportunities for collaboration, referrals, or business development. This might include potential clients, referral partners, industry peers, or thought leaders in your field.

Use LinkedIn's advanced search features to find professionals based on specific criteria including industry, company size, geographic location, seniority level, and mutual connections. These targeted searches help you identify high-value connection opportunities.

Personalize every connection request with specific reasons for connecting and potential value you could provide. Generic connection requests get ignored, while thoughtful messages that demonstrate genuine interest and potential mutual benefit generate positive responses.

Follow up on accepted connections with valuable content, helpful introductions, or relevant opportunities rather than immediate sales pitches. Relationship building takes time and consistent value delivery before business opportunities naturally emerge.

LinkedIn Groups and Community Engagement

Participate actively in LinkedIn groups where your ideal clients and professional peers gather to discuss industry topics and challenges. Group participation positions you as an expert while building relationships with potential clients and referral sources.

Share valuable insights and answer questions in group discussions rather than promoting your services directly. Helpful contributions build your reputation within the community and encourage private messages and connection requests from interested members.

Start meaningful discussions by posting thoughtful questions or sharing industry insights that generate engagement and position you as a thought leader within the group community.

Use group membership lists to identify potential connections who are already engaged in your professional community and likely to be interested in building relationships with relevant experts.

Lead Generation and Business Development

Use LinkedIn Sales Navigator or advanced search features to identify potential clients based on specific criteria relevant to your services. Research their challenges, recent company developments, and mutual connections that could facilitate introductions.

Engage with prospects' content before reaching out directly. Like, comment meaningfully on, and share their posts to build familiarity and demonstrate genuine interest in their work and perspectives.

Craft personalized outreach messages that focus on their challenges and how your expertise might help rather than immediately pitching your services. The goal is to start conversations that could develop into business relationships over time.

Track your networking activities and follow up systematically with connections who have expressed interest in your services. Use LinkedIn's note-taking features to remember conversation details and maintain personalized communication.

Action Steps for Chapter 4:

Action Item	Task Description	Expected Outcome
LinkedIn Profile Optimization	Complete a comprehensive profile update including compelling headline, detailed About section, optimized experience descriptions, and strategic skills selection. Include professional photos and contact information.	Professional LinkedIn presence that effectively communicates your value and attracts your ideal audience
Content Publishing Strategy	Develop a content calendar for LinkedIn posting including 20 initial post ideas across different content types (industry insights, case studies, professional development). Commit to posting 3-5 times per week.	Consistent thought leadership presence that builds authority and engagement with your target audience
Strategic Connection Campaign	Identify 100 potential connections in your target market and send personalized connection requests. Follow up with accepted connections using value-first approaches rather than sales pitches.	Expanded professional network of relevant contacts who could become clients, collaborators, or referral sources

Action Item	Task Description	Expected Outcome
Group Engagement Program	Join 5-10 LinkedIn groups where your ideal clients gather and commit to participating actively by answering questions, sharing insights, and starting valuable discussions.	Increased visibility and credibility within your professional community leading to connection requests and business opportunities
Lead Generation System	Set up a systematic approach for identifying and reaching out to potential clients on LinkedIn. Create templates for outreach messages and establish a follow-up schedule for nurturing relationships.	Predictable pipeline of qualified prospects generated through strategic LinkedIn networking and relationship building

FIVE

Becoming a Voice on X (Twitter)

X (formerly Twitter) represents the global town square for real-time professional discourse, breaking news, and thought leadership across every industry and discipline. The platform's character limit forces concise, impactful communication that cuts through noise and delivers immediate value to busy professionals seeking quick insights and timely perspectives.

Building influence on X requires mastering the art of distilling complex ideas into memorable, shareable insights that demonstrate expertise while sparking meaningful conversations. The most successful thought leaders on the platform combine deep knowledge with accessible communication, making sophisticated concepts understandable and actionable for broad audiences.

Unlike other platforms that favor polished content, X rewards authenticity, timeliness, and genuine engagement with current events and industry developments. Your influence grows through consistent valuable contributions to ongoing conversations rather than broadcasting promotional messages.

The Art of Concise Thought Leadership

Effective X thought leadership transforms complex professional insights into clear, actionable statements that busy professionals can quickly understand and apply. This requires understanding your core message and expressing it with precision and impact.

Start with comprehensive knowledge and distill it down to essential elements that provide immediate value. If you understand a complex business process, identify the three most critical success factors. If you've learned from a major failure, extract the key lesson that others can apply to avoid similar mistakes.

Use specific examples and concrete language rather than abstract concepts. "Successful product launches require clear success metrics, realistic timelines, and cross-functional alignment" is more valuable than "Product management involves strategic planning and execution excellence."

Test different ways of expressing the same insights to identify language and framing that generates the most engagement and resonates most strongly with your target audience. The same principle might be expressed in multiple ways that appeal to different professional perspectives.

Voice Development for Professional Impact

Your X voice should reflect your authentic professional personality while adapting to the platform's conversational, immediate culture. The best professional voices on X combine expertise with approachability, creating content that feels both authoritative and relatable.

Develop signature phrases, perspectives, or frameworks that become associated with your brand. This might be a particular way of analyzing industry trends, a consistent methodology you apply to problems, or a unique perspective that differentiates your insights from others in your field.

Balance professional insights with personality elements that make your content memorable and shareable. Humor, personal anecdotes, and candid observations can enhance rather than detract from your professional credibility when used appropriately.

Maintain consistency in your core values and perspectives while adapting your communication style to different topics and audiences. Your fundamental expertise and worldview should remain recognizable across all your content.

Timing and Relevance Strategies

X rewards timeliness and relevance more than other platforms, making it essential to stay current with industry developments, news events, and trending topics that affect your professional community.

Monitor industry news, company announcements, and economic developments that create opportunities for valuable commentary from your expertise perspective. Quick, insightful responses to breaking news can generate significant visibility and engagement.

Participate in industry conversations while they're happening rather than commenting after interest has moved elsewhere. Set up alerts for keywords, company names, and industry terms that might generate opportunities for valuable contribution.

Connect current events to timeless professional principles and insights. When news breaks in your industry, help others understand the broader implications, lessons learned, or strategic considerations that extend beyond the immediate story.

THREADS AND VIRAL STRATEGIES

X threads allow for longer-form storytelling and comprehensive insight sharing while maintaining the platform's conversational feel and engaging format. Well-crafted threads can go viral, dramatically expanding your reach and establishing your expertise with new audiences who discover your content through shares and engagement.

Thread creation requires understanding both storytelling principles and X's unique culture and algorithm preferences. The most successful threads combine valuable information with an engaging narrative structure that keeps readers clicking through to the conclusion.

Viral content on X typically provides either immediate practical value or strong emotional resonance, often combining both elements to create content that people feel compelled to share with their networks.

Thread Structure and Storytelling

Begin threads with compelling hooks that promise specific value or intriguing insights. Your opening tweet should clearly communicate what readers will gain by investing time in your full thread.

Structure threads like engaging stories with clear beginning, middle, and end sections. Even educational or analytical threads benefit from narrative progression that maintains interest throughout the entire sequence.

Use numbered tweets to help readers follow the logical progression and know how much content remains. This structure also makes threads easier to bookmark and reference later.

Include visual elements, data, or examples that break up text-heavy content and provide additional value. Screenshots, charts, or infographics can significantly increase engagement and shareability.

Conclude threads with clear takeaways and calls to action that encourage engagement, follows, or further discussion. Strong endings often determine whether readers share content with their networks.

Viral Content Principles

Viral X content typically addresses universal professional challenges, provides actionable insights, or offers unique perspectives on widely discussed topics. The key is finding fresh angles on subjects that already have audience interest and engagement.

Educational threads that teach valuable skills or share hard-won lessons tend to perform well because they provide immediate practical value that readers want to save and share with colleagues.

Contrarian takes on popular opinions can generate significant engagement and discussion, but ensure your perspective is well-reasoned and adds genuine value to the conversation rather than being controversial for attention alone.

Personal stories that illustrate broader professional principles often resonate strongly because they combine relatability with actionable insights. Share failures, learning experiences, and breakthrough moments that others can learn from.

Engagement and Community Building

Respond actively to comments and questions on your threads to encourage continued discussion and signal to the algorithm that your content is generating valuable engagement.

Engage with others who share or comment on your threads by visiting their profiles, following valuable contributors, and participating in their content discussions. This reciprocal engagement builds relationships and expands your network.

Use thread success to identify topics and formats that resonate most strongly with your audience, then create additional content around similar themes while avoiding exact repetition.

Track which threads generate the most profile visits, follows, and business inquiries to understand what content best serves your professional objectives beyond just engagement metrics.

ENGAGING IN INDUSTRY CONVERSATIONS FOR VISIBILITY

X's real-time nature makes it the primary platform for industry discussions, breaking news commentary, and professional debate across virtually every field. Active participation in these conversations positions you as an engaged expert while building relationships with other industry voices and potential clients or collaborators.

Successful industry engagement requires balance between contributing valuable insights and avoiding controversial topics that could damage your professional reputation. The goal is to become known as a thoughtful contributor whose perspectives add value to ongoing discussions.

Strategic conversation participation builds your network, increases your visibility, and establishes you as someone worth following for industry insights and professional development.

Identifying High-Value Conversations

Monitor hashtags, trending topics, and ongoing discussions relevant to your industry to identify opportunities for valuable contribution. Use X's search functions and lists to track conversations among industry leaders and potential clients.

Follow industry publications, thought leaders, and companies whose content regularly generates meaningful professional discussions. These accounts often serve as conversation starters that create opportunities for valuable engagement.

Set up keyword alerts for terms related to your expertise, industry developments, or client challenges that might create opportunities for helpful responses or insightful commentary.

Participate in recurring industry conversations like weekly hashtag discussions, live-tweeting industry events, or commenting on quarterly earnings reports that affect your professional community.

Strategic Engagement Approaches

Add substantive value to conversations rather than simply agreeing or sharing obvious observations. Provide additional context, alternative perspectives, or practical applications that enhance the original discussion.

Ask thoughtful questions that advance conversations and demonstrate your expertise while encouraging others to share their perspectives and experiences.

Share relevant resources, tools, or insights that help others in the conversation achieve better outcomes or understand complex topics more clearly.

Acknowledge others' contributions and build on their insights rather than contradicting or dismissing different viewpoints, which builds relationships while maintaining professional credibility.

Building Professional Relationships

Use industry conversations as opportunities to identify and connect with other professionals whose work and perspectives you respect and who might become valuable network connections.

Follow up on meaningful conversation exchanges with direct messages or connection requests that reference your interaction and suggest potential mutual value.

Offer to help others in conversations when you have relevant expertise or resources that could benefit them, building goodwill and professional relationships.

Share and amplify others' valuable content with thoughtful commentary that adds to their insights while building reciprocal relationships with industry peers.

Monitoring and Managing Your Professional Reputation

Regularly review your X activity to ensure your professional reputation remains positive and your contributions consistently add value to industry discussions.

Avoid engaging in heated debates or controversial topics that could alienate potential clients or damage your professional reputation, even when you have strong opinions about industry issues.

Focus on being helpful, insightful, and professional in all your interactions, remembering that potential clients and collaborators may review your X history when evaluating whether to work with you.

Use X's analytics to track which types of engagement generate the most positive responses and business inquiries, focusing your conversation participation on the most professionally valuable activities.

Action Steps for Chapter 5:

Action Item	Task Description	Expected Outcome
X Profile and Voice Development	Optimize your X profile with professional photo, compelling bio, and relevant links. Develop your unique voice by posting 10–15 tweets that demonstrate your expertise and personality, testing different tones and topics.	Established X presence with clear professional positioning and authentic voice that resonates with your target audience
Thread Content Creation	Plan and publish 5 high-value threads covering different aspects of your expertise. Use storytelling structure with compelling hooks, valuable insights, and clear calls-to-action. Monitor performance to identify successful formats.	Demonstrated thought leadership through comprehensive content that showcases expertise while building audience engagement

Action Item	Task Description	Expected Outcome
Industry Conversation Mapping	Identify 20 industry leaders, publications, and hashtags to follow for conversation opportunities. Set up keyword alerts and create lists to monitor relevant discussions in your professional area.	Systematic approach to discovering and participating in high-value professional conversations that build visibility and credibility
Daily Engagement Strategy	Establish a daily routine of posting valuable content, responding to industry discussions, and engaging with other professionals' content. Aim for 5–10 meaningful interactions per day beyond your own posting.	Consistent presence in industry conversations that builds relationships and establishes you as an active, valuable community member
Performance and Reputation Monitoring	Set up tracking for your X analytics, monitoring engagement rates, profile visits, and follower growth. Regularly review your content for professional consistency and reputation impact.	Data-driven understanding of what content and engagement strategies generate the best professional outcomes and business opportunities

SIX

Visual Storytelling on Instagram

Instagram's visual-first platform requires a fundamentally different approach to personal branding than text-based platforms. Success on Instagram means thinking like a visual storyteller who uses images, videos, and design elements to communicate expertise, personality, and professional values in ways that create emotional connections with your audience.

The most effective professional Instagram brands balance aspirational lifestyle content with authentic behind-the-scenes glimpses that humanize expertise while maintaining credibility. Your visual choices become extensions of your professional brand, communicating values, personality, and quality standards before followers read a single caption.

Visual storytelling on Instagram isn't about perfection—it's about consistency, authenticity, and strategic curation that supports your professional objectives while creating content that people genuinely want to follow and engage with regularly.

Developing Your Visual Brand Identity

Your Instagram visual brand should feel cohesive and intentional while remaining authentic to your actual lifestyle and professional context. This means developing consistent approaches to photography, color schemes, and content themes that reflect both your expertise and personality.

Color consistency creates instant recognition and professional polish. This doesn't mean every image must use identical colors, but establishing a palette of complementary colors that appear regularly throughout your content creates visual cohesion that makes your posts immediately recognizable.

Photography style consistency includes lighting preferences, composition approaches, and editing techniques that create a unified look across all your content. Whether you prefer bright, clean aesthetics or moody, dramatic lighting, consistency in your approach builds brand recognition.

Content theme balance involves mixing professional expertise with personal elements that make you relatable and interesting to follow. Business coaches might share workspace organization alongside family moments that demonstrate work-life integration, while consultants might combine client meeting insights with travel content that shows their mobile lifestyle.

Establish visual guidelines for yourself including preferred photo editing approaches, color schemes, and composition styles that you can reference when creating content. This ensures consistency while allowing creative flexibility within your established parameters.

Professional Lifestyle Integration

Instagram audiences want to see the person behind the professional brand, which requires thoughtful integration of personal and professional content that supports rather than undermines your credibility and expertise.

Behind-the-scenes content that shows your work process, preparation routines, or professional environment helps followers understand how you operate while maintaining appropriate professional boundaries. This might include workspace setups, preparation for client meetings, or the resources you use for professional development.

Personal elements that reinforce your professional values create authentic connections without compromising credibility. If you help busy professionals achieve better work-life balance, sharing content about your own family time, exercise routines, or relaxation practices demonstrates authentic commitment to the principles you teach.

Travel and location content can showcase the freedom and lifestyle that professional success enables while reinforcing your expertise. Remote work setups, client meetings in interesting locations, or speaking engagements in different cities all reinforce professional credibility while providing engaging visual content.

Essential Instagram Visual Brand Elements:

- **Consistent Filter/Editing Style**: Develop a signature photo editing approach using specific apps or presets that create unified visual aesthetics. Whether you prefer bright and airy, warm and cozy, or bold and dramatic, consistency in editing creates instant brand recognition.
- **Template Design System**: Create branded templates for quote graphics, educational carousels, announcement posts, and story highlights that maintain visual consistency while allowing efficient content creation. Use consistent fonts, colors, and layouts across all designed elements.
- **Grid Planning Strategy**: Plan your Instagram grid layout to create visual flow and balance between different content types. Use planning apps to preview how posts will look together and ensure your profile creates a strong first impression for new visitors.

- **Story Highlight Covers**: Design custom covers for your Instagram story highlights that match your brand aesthetic and clearly communicate the content themes. These covers become part of your profile's visual identity and help organize your content for easy navigation.
- **Professional Photography Standards**: Maintain consistent quality standards for all photos including proper lighting, sharp focus, and thoughtful composition. Invest in basic photography equipment or apps that help you create professional-looking content consistently.
- **Brand Color Integration**: Incorporate your brand colors naturally throughout your content through props, backgrounds, clothing choices, or graphic elements. This subtle integration reinforces brand recognition without appearing forced or unnatural.

REELS, STORIES, AND CAROUSELS THAT BUILD TRUST AND RELATABILITY

Instagram's diverse content formats provide multiple opportunities to showcase different aspects of your expertise and personality. Reels capture attention through entertaining or educational short-form video content, Stories build daily connection through casual, temporary content, and carousels allow for comprehensive education through multi-slide posts.

Each format serves different purposes in your overall brand strategy and appeals to different content consumption preferences among your followers. Mastering all three formats creates a comprehensive Instagram presence that engages various audience segments while reinforcing your professional brand through different content experiences.

The key to success with Instagram content formats is understanding how each one contributes to your larger professional objectives while providing genuine value that keeps followers engaged and attracts new audience members.

Instagram Reels for Professional Engagement

Reels provide powerful opportunities to demonstrate expertise through quick tutorials, industry insights, or behind-the-scenes professional content that entertains while educating your audience about your field.

Educational Reels that teach specific skills or share valuable insights often perform well because they provide immediate practical value. Break complex professional concepts into simple, actionable steps that viewers can implement immediately.

Behind-the-scenes Reels showing your work process, daily routines, or professional preparation create authentic connections while demonstrating your expertise in action. These might include client meeting preparation, content creation processes, or professional development activities.

Trend participation adapted to your professional context can increase visibility while maintaining brand relevance. Use trending audio or visual effects to create content that feels current and engaging while showcasing your expertise.

Industry commentary Reels that provide your perspective on current events, trends, or developments in your field position you as a thought leader while creating content that encourages discussion and engagement.

Stories for Daily Connection and Engagement

Instagram Stories provide opportunities for more casual, authentic content that builds personal connections with your audience through daily glimpses into your professional and personal life.

Professional updates about client successes, project completions, or business milestones shared through Stories feel more personal and authentic than formal posts while still reinforcing your expertise and success.

Q&A sessions using Instagram's question feature allow direct interaction with your audience while demonstrating your knowledge and willingness to help others with their challenges.

Behind-the-scenes Story content including preparation for client meetings, workspace organization, or professional development activities provides authentic glimpses into your work life without requiring polished presentation.

Real-time industry commentary on news, trends, or events relevant to your field shows that you stay current with industry developments and have valuable perspectives to share.

Carousel Posts for Comprehensive Education

Carousel posts allow for detailed educational content that demonstrates expertise while providing significant value to your audience. These multi-slide posts work well for step-by-step processes, comprehensive guides, or detailed case studies.

Educational carousels that break complex topics into digestible slides perform well because they provide substantial value while being easy to save and reference later. Focus on actionable insights that readers can implement immediately.

Case study carousels showcasing client results or project outcomes provide social proof while demonstrating your methodology and expertise in action. Include specific results and lessons learned that others can apply to their situations.

List-based carousels that compile resources, tools, tips, or strategies related to your expertise area provide immediate practical value while positioning you as a knowledgeable resource in your field.

Quote and insight carousels that present your professional philosophy, key principles, or important lessons learned help establish thought leadership while creating highly shareable content.

CONVERTING FOLLOWERS INTO LOYAL FANS OR CLIENTS

Instagram's visual nature creates strong emotional connections that can translate into business relationships when nurtured strategically through valuable content, authentic engagement, and clear communication about how you help your audience achieve their goals.

Converting Instagram followers into clients requires building trust through consistent value delivery, demonstrating expertise through helpful content, and creating natural pathways for interested followers to learn more about your services and potentially engage professionally.

The conversion process on Instagram typically happens gradually through relationship building rather than direct sales approaches. Your content should consistently provide value while subtly demonstrating your expertise and the results you help others achieve.

Building Trust Through Consistent Value

Establish yourself as a reliable source of valuable information by consistently sharing insights, tips, strategies, and resources that help your audience achieve their professional or personal goals.

Educational content that solves real problems for your audience builds trust while demonstrating your expertise and ability to help others achieve better outcomes in areas where you specialize.

Behind-the-scenes content that shows your work process, client interactions (with appropriate permissions), or professional development activities builds authenticity and trust by showing the real person behind the professional brand.

Client success stories and testimonials shared through various content formats provide social proof while demonstrating your ability to generate results for others facing similar challenges.

Consistent posting schedules and reliable content quality create expectations that build trust over time. Followers who know they can count on you for valuable content are more likely to consider your professional services when needs arise.

Strategic Call-to-Actions and Lead Generation

Use Instagram's link-in-bio feature to direct interested followers to lead magnets, consultation bookings, or comprehensive information about your services. Update this link regularly to align with current business objectives.

Story features like link stickers, swipe-ups (for accounts with 10k+ followers), and call-to-action buttons provide opportunities to direct engaged viewers to specific actions that support your business goals.

Direct message conversations often provide the most natural pathway from social media follower to business inquiry. Encourage DMs through your content and respond promptly with helpful information and clear next steps.

Email list building through valuable lead magnets promoted on Instagram creates owned marketing channels that allow deeper relationship building and more direct communication about your services.

Nurturing Long-term Relationships

Respond personally to comments and direct messages to build individual relationships with engaged followers who might become clients or valuable referral sources over time.

Share user-generated content and highlight follower successes to create community feeling while demonstrating the positive impact of your expertise and content.

Use Instagram's various features including Stories, Live videos, and IGTV to create different types of engagement opportunities that appeal to different followers' preferences and deepen relationships over time.

Track which content types and engagement strategies lead to the most business inquiries and focus your efforts on activities that generate qualified prospects rather than just general engagement.

Action Steps for Chapter 6:

Action Item	Task Description	Expected Outcome
Visual Brand Development	Establish your Instagram visual identity including color palette, photography style, and editing approach. Create templates for common content types and plan your grid layout strategy for visual consistency.	Cohesive Instagram brand that creates immediate recognition and professional credibility through consistent visual elements
Content Format Mastery	Create 10 pieces of content using different Instagram formats: 3 educational Reels, 3 value-packed carousels, daily Stories for one week, and regular feed posts. Analyze performance to identify most effective formats.	Expertise in all Instagram content formats with data-driven understanding of what resonates most with your specific audience
Engagement and Community Building	Develop a daily engagement routine including responding to comments, engaging with followers' content, and using Stories features for direct interaction. Aim for authentic connection rather than just promotional activity.	Strong community relationships that build trust and increase likelihood of converting followers into clients or advocates

Action Item	Task Description	Expected Outcome
Lead Generation System	Set up Instagram lead generation including optimized bio with clear value proposition, link-in-bio strategy, lead magnets promoted through Stories and posts, and systems for capturing and nurturing interested prospects.	Systematic approach to converting Instagram engagement into business inquiries and qualified prospects
Content Performance Analysis	Track Instagram analytics to identify top-performing content types, optimal posting times, and engagement patterns. Use insights to refine content strategy and focus on activities that drive both engagement and business results.	Data-driven Instagram strategy that maximizes both audience growth and business development opportunities

SEVEN

Long-Form Influence on YouTube

YouTube represents the world's second-largest search engine and serves as the premier platform for establishing long-form thought leadership and comprehensive expertise demonstration. Unlike other social platforms that prioritize brief interactions, YouTube rewards depth, teaching ability, and sustained value delivery that builds lasting authority in your professional field.

Your YouTube channel functions as your digital university, providing space to share comprehensive knowledge, detailed case studies, and thorough explanations that establish you as the definitive expert in your niche. This depth of content creates stronger authority than any other platform while building an archive of valuable resources that attract new audiences for years after publication.

Building authority on YouTube requires strategic channel positioning that clearly communicates your unique expertise while creating content systems that consistently deliver exceptional value to your target audience. Your channel should become the destination where professionals in your field come for the most comprehensive, reliable information available.

Channel Strategy and Positioning

Define your channel's specific focus and value proposition with precision that differentiates you from the thousands of other professionals creating content in your general field. Rather than being "a marketing expert," position yourself as "the authority on email marketing for B2B SaaS companies" or "the go-to resource for productivity systems for creative entrepreneurs."

Your channel trailer and about section should immediately communicate who you serve, what specific problems you solve, and what unique perspective or methodology you bring to your field. New visitors should understand within seconds whether your channel is relevant to their needs and interests.

Develop content pillars that establish your expertise across the key areas where your target audience needs guidance. These pillars become recurring themes that demonstrate comprehensive knowledge while providing systematic coverage of your field.

Create a consistent publishing schedule that builds audience expectations and algorithmic favor. Whether you publish weekly or monthly, consistency in timing and quality builds trust while supporting YouTube's algorithm preferences for reliable content creators.

Authority-Building Content Types

Comprehensive tutorials and educational series that teach complex skills or processes in detail demonstrate your expertise while providing significant value that audiences bookmark and return to repeatedly. These in-depth resources become your channel's foundation content.

Case study analyses that break down successful projects, campaigns, or strategies provide behind-the-scenes insights that most professionals never share publicly. This transparency builds trust while demonstrating your analytical capabilities and real-world experience.

Industry analysis and commentary videos that interpret trends, developments, or news relevant to your field position you as someone who understands the broader context and can help others navigate change and opportunity.

Q&A sessions and response videos that address audience questions demonstrate responsiveness while covering topics that your audience specifically wants to learn about, ensuring content relevance and engagement.

YouTube Authority Development Framework:

1. **Expertise Documentation Strategy**: Create comprehensive content that documents your methodologies, processes, and frameworks in detail. This might include step-by-step guides for your signature approaches, detailed breakdowns of successful projects, or comprehensive explanations of industry concepts that establish your knowledge depth.
2. **Thought Leadership Positioning**: Develop unique perspectives, frameworks, or methodologies that become associated with your personal brand. Share contrarian viewpoints supported by evidence, create new terminology or concepts that advance your field, or synthesize existing knowledge in novel ways that provide fresh insights.
3. **Teaching Ability Demonstration**: Focus on creating content that actually helps viewers achieve specific outcomes rather than just sharing information. The ability to teach effectively—to transform complex concepts into actionable knowledge—establishes deeper authority than simply having expertise.
4. **Consistency and Reliability Building**: Maintain regular publishing schedules and consistent quality standards that build audience trust and algorithmic support. Establish yourself as someone your audience can count on for reliable, valuable content that consistently meets their professional development needs.

BALANCING ENTERTAINMENT WITH EDUCATION

Successful YouTube content combines educational value with engaging presentation that holds viewer attention throughout longer-form content. Unlike text-based platforms where information alone might suffice, YouTube's video format requires understanding of pacing, storytelling, and visual engagement that keeps audiences watching until completion.

The most successful professional YouTube channels understand that education and entertainment aren't opposing forces—they're complementary elements that work together to create memorable, impactful content that audiences eagerly anticipate and readily share with colleagues and networks.

Balancing these elements requires understanding your audience's learning preferences while developing presentation skills that make complex information accessible and engaging for video consumption.

Educational Content Structure

Structure educational videos with clear learning objectives stated upfront so viewers understand the specific value they'll gain by watching to completion. This expectation setting improves retention while helping you deliver on promised outcomes.

Break complex topics into logical segments with clear transitions that help viewers follow your reasoning and retain information effectively. Use verbal signposts like "first," "next," and "finally" to guide viewers through your content structure.

Include practical examples, case studies, or demonstrations that illustrate abstract concepts with concrete applications viewers can relate to their own situations and challenges.

Provide actionable takeaways and next steps that viewers can implement immediately after watching your video. This immediate applicability increases content value while encouraging viewers to return for additional guidance.

Entertainment Elements That Enhance Learning

Storytelling techniques, including personal anecdotes, client experiences, or case study narratives, make educational content more memorable while demonstrating real-world application of the concepts you're teaching.

Visual variety through screen sharing, graphics, props, or location changes maintains viewer interest during longer videos while supporting different learning styles and information processing preferences.

Personality integration through humor, enthusiasm, or authentic reactions to your content creates emotional connection while making educational material more enjoyable to consume.

Pacing variation that includes faster-paced segments for energy and slower explanations for complex concepts keeps viewers engaged while ensuring comprehension of important information.

Audience Engagement Optimization

Interactive elements like questions, polls, or calls-to-action encourage active participation rather than passive consumption, increasing engagement while providing valuable feedback about audience interests and needs.

Community building through consistent interaction with comments, addressing viewer questions in future videos, and acknowledging regular viewers by name creates loyalty while encouraging continued engagement.

Series development that creates anticipation for upcoming content keeps audiences returning while allowing comprehensive coverage of complex topics that require multiple videos to address completely.

Feedback integration that responds to viewer suggestions, questions, and requests demonstrates responsiveness while ensuring your content remains relevant to audience needs and interests.

MONETIZATION THROUGH ADS, SPONSORSHIPS, AND PRODUCTS

YouTube's mature monetization ecosystem provides multiple revenue streams that can generate significant income for channels that build engaged audiences and maintain consistent content creation. Understanding these monetization options allows strategic content creation that optimizes for both audience value and revenue generation.

Successful YouTube monetization requires building sufficient audience size and engagement to qualify for various programs while creating content that naturally integrates promotional elements without compromising educational value or audience trust.

The most sustainable YouTube revenue comes from diversified income streams that reduce dependence on any single monetization method while maximizing the value extracted from your content creation efforts and audience relationships.

YouTube Partner Program Optimization

YouTube's Partner Program provides ad revenue sharing that requires 1,000 subscribers and 4,000 watch hours in the past 12 months. Meeting these thresholds requires strategic content creation focused on audience building and engagement optimization.

Ad revenue optimization involves understanding YouTube's advertiser-friendly content guidelines while creating videos that attract premium advertising rates through professional presentation and valuable content that appeals to business audiences.

Video length strategy affects ad placement opportunities—videos longer than 8 minutes can include mid-roll ads that significantly increase revenue potential compared to shorter videos limited to pre-roll advertisements.

Audience demographics influence ad rates because advertisers pay premium prices for specific audience segments. Business and professional development content often attracts higher advertising rates than entertainment-focused channels.

Sponsorship and Brand Partnership Development

Direct sponsorships often provide more revenue than ad sharing while allowing greater control over promotional content integration and brand alignment with your professional values and audience interests.

Media kit development that showcases your audience demographics, engagement statistics, and previous collaboration examples helps attract brand partnerships while demonstrating professionalism that justifies premium sponsorship rates.

Integration strategies that naturally incorporate sponsored content into valuable educational material maintain audience trust while fulfilling sponsor objectives for authentic product or service endorsement.

Long-term partnership cultivation with brands that align with your audience's needs and interests creates ongoing revenue streams while reducing the effort required to continuously seek new sponsorship opportunities.

Product and Service Promotion

YouTube serves as powerful marketing for your own products and services through demonstration, education, and trust building that naturally leads to business inquiries and client relationships.

Course creation and promotion through YouTube content that provides sample teaching and demonstrates your expertise can generate significant revenue while serving audiences who want comprehensive education.

Consulting and coaching promotion through case studies, methodology explanations, and client success stories builds credibility while attracting high-value professional service inquiries.

Affiliate marketing integration for tools, resources, or services you genuinely use and recommend provides additional revenue while helping your audience access valuable resources relevant to your educational content.

Revenue Stream Integration Strategy

Diversified monetization approaches reduce risk while maximizing income potential from your content creation efforts. Combine ad revenue, sponsorships, product sales, and service promotion strategically.

Content calendar planning that balances educational value with monetization opportunities ensures audience trust while optimizing revenue generation through strategic promotional integration.

Analytics tracking that monitors which monetization methods generate the best results allows optimization of your revenue strategy while maintaining content quality and audience satisfaction.

Long-term strategy development that considers how different revenue streams support business growth while building sustainable income that justifies continued content creation investment.

Action Steps for Chapter 7:

Action Item	Task Description	Expected Outcome
YouTube Channel Setup and Optimization	Create or optimize your YouTube channel with professional branding, clear value proposition, compelling channel trailer, and organized playlists. Develop content pillars that establish your expertise across key topic areas.	Professional YouTube presence positioned as the authority hub for your specific expertise area with clear value communication
Content Creation and Publishing System	Plan and create your first 10 YouTube videos using educational content structure with entertainment elements. Establish a consistent publishing schedule and develop efficient production workflows.	Systematic approach to creating engaging, valuable content that builds authority while maintaining sustainable production processes
Audience Engagement Strategy	Implement community building practices including responding to comments, creating interactive content, asking for viewer feedback, and developing series that encourage return viewership.	Engaged YouTube community that provides feedback, shares content, and becomes a source of business opportunities and referrals
Monetization Planning	Research and plan your YouTube monetization strategy including Partner Program preparation, potential sponsorship targets, and integration of your own products or services. Create a media kit for brand partnerships.	Clear monetization roadmap that maximizes revenue potential while maintaining audience trust and content quality
Performance Analytics and Optimization	Set up YouTube analytics tracking for key metrics including watch time, subscriber growth, engagement rates, and revenue performance. Create a monthly review process for strategy optimization.	Data-driven approach to YouTube growth that identifies successful content types and monetization strategies for continued optimization

EIGHT

Community and Trust on Facebook

Facebook's mature platform provides unique opportunities for building meaningful professional communities through Groups and establishing business credibility through Pages. Unlike newer social platforms focused on discovery and viral content, Facebook excels at fostering ongoing relationships and facilitating deep community engagement around shared interests and professional development.

Facebook Groups create intimate spaces where professionals can share challenges, seek advice, and build relationships that often translate into business opportunities, referrals, and long-term professional collaborations. The platform's sophisticated group management tools allow for moderated discussions, exclusive content, and community building that develops genuine trust and mutual support.

Building successful Facebook communities requires understanding the platform's emphasis on meaningful social interaction and authentic relationship development rather than broadcasting or promotional activity. The most valuable Facebook presence focuses on serving others and facilitating valuable connections rather than self-promotion.

Strategic Group Creation and Management

Creating a Facebook Group around your area of expertise establishes you as a community leader while providing a platform for ongoing relationship building with potential clients, collaborators, and referral sources. Successful groups focus on specific niches rather than broad professional categories.

Choose group topics that address genuine ongoing needs in your professional community rather than topics that generate initial interest but lack sustainable engagement. "Marketing for Local Service Businesses" provides more focused value than "Business Marketing Tips."

Establish clear group guidelines that encourage helpful discussion while maintaining professional standards and preventing spam or inappropriate content. Well-managed groups attract higher-quality members who become valuable network connections.

Facilitate regular engagement through discussion prompts, expert interviews, resource sharing, and member spotlights that create value for participants while positioning you as a knowledgeable facilitator and connector.

Facebook Page Professional Development

Facebook Pages serve as your official business presence on the platform, providing credibility and comprehensive information about your professional services while integrating with Facebook's advertising and messaging systems.

Optimize your Page with complete business information, professional photography, clear service descriptions, and regular content that demonstrates expertise and builds trust with potential clients who research your background.

Use Page features like Services sections, Reviews, and Contact information to make it easy for potential clients to understand your offerings and initiate business conversations.

Maintain consistent posting schedules and professional presentation that reinforces your credibility while providing value to followers who use your Page to stay updated on your professional activities.

Community Engagement Best Practices:

- **Value-First Participation**: Focus on helping others solve problems, sharing valuable resources, and facilitating connections rather than promoting your services directly. Consistent value delivery builds trust that naturally leads to business opportunities.
- **Expert Commentary**: Provide thoughtful analysis and insights on industry developments, trending topics, and member questions that demonstrate your knowledge depth while helping others understand complex issues.
- **Resource Curation**: Share valuable tools, articles, studies, and resources that benefit group members while positioning yourself as someone who stays current with industry developments and helpful resources.
- **Connection Facilitation**: Introduce group members to each other when you identify potential collaboration opportunities, helping build community relationships while establishing yourself as a valuable network connector.
- **Content Moderation**: Maintain group quality by enforcing guidelines consistently, removing spam or inappropriate content, and guiding discussions toward productive outcomes that serve member needs.
- **Member Recognition**: Highlight member achievements, successful implementations of shared advice, and valuable contributions to create positive community culture and encourage ongoing participation.
- **Educational Series**: Create ongoing educational content series within groups that demonstrate expertise while providing systematic value that keeps members engaged and returning regularly.

- **Live Interaction**: Use Facebook Live features to host Q&A sessions, educational workshops, or informal discussions that create real-time connection and demonstrate personality alongside professional expertise.

LIVE SESSIONS FOR RELATIONSHIP BUILDING

Facebook Live provides powerful opportunities for real-time connection and authentic relationship building that creates stronger professional bonds than text-based interaction alone. Live video allows your audience to experience your personality, expertise, and communication style in ways that build trust and encourage business relationships.

The platform's Live features include audience interaction capabilities that create genuine two-way communication during broadcasts. This immediate feedback and personal recognition helps individual community members feel valued while building stronger overall group cohesion and engagement.

Successful Facebook Live sessions balance planned content with spontaneous interaction, creating valuable educational experiences while allowing personality and authenticity to build deeper professional relationships than polished, scripted content could achieve.

Planning Effective Live Sessions

Structure Live sessions with clear objectives and valuable content that justifies audience time investment while leaving space for spontaneous interaction and question response that creates personal connection.

Educational Live sessions that teach specific skills, share case studies, or provide industry analysis offer immediate value while demonstrating your expertise in real-time application and teaching ability.

Q&A format sessions allow direct interaction with your audience while addressing their specific challenges and questions, providing personalized value while showcasing your knowledge breadth and problem-solving approach.

Behind-the-scenes Live content that shares your work process, daily routines, or professional insights creates authenticity while demonstrating your expertise in natural, unscripted contexts.

Audience Interaction and Engagement

Acknowledge viewers by name during Live sessions to create personal connection while encouraging continued participation and return viewership for future broadcasts.

Respond to comments and questions in real-time to demonstrate accessibility and genuine interest in helping your audience, building trust that often translates into business inquiries.

Use Facebook's Live features like polls, reactions, and comment prompts to encourage active participation rather than passive viewing, creating more engaging experiences that build stronger community connections.

Follow up after Live sessions with participants who asked questions or showed strong engagement, using direct messages or additional content to continue conversations and build individual relationships.

Building Professional Authority Through Live Content

Demonstrate expertise through impromptu problem-solving during Live sessions, showing your ability to think on your feet while providing valuable guidance to real audience challenges.

Share case studies and real examples during Live sessions to illustrate concepts with concrete applications that help audiences understand how your expertise applies to their situations.

Address industry developments or current events during Live sessions to position yourself as someone who stays current with important trends while helping others understand implications and opportunities.

Create recurring Live session themes or series that build audience expectations while systematically covering important topics in your field, establishing comprehensive thought leadership over time.

USING FACEBOOK AS A FUNNEL FOR BUSINESS GROWTH

Facebook's sophisticated targeting capabilities and established user base make it an excellent platform for systematically building business funnels that move prospects from initial awareness through relationship development to client conversion. The platform's mature advertising system and organic reach opportunities work together to create comprehensive business development strategies.

Unlike newer platforms where business integration might feel forced, Facebook's users expect and accept business-oriented content when it provides genuine value. This acceptance allows for more direct business development approaches while maintaining focus on relationship building and value delivery.

Successful Facebook business funnels combine organic community building with strategic advertising to create systematic lead generation and client development processes that generate consistent business opportunities over time.

Lead Generation and List Building

Facebook's lead generation ads allow direct email capture within the platform, reducing friction while building email lists of prospects interested in your expertise and services.

Create valuable lead magnets specifically designed for Facebook audiences, considering the platform's demographics and user preferences for comprehensive resources and detailed information.

Use Facebook Groups to naturally promote lead magnets through helpful content that references additional resources available through email signup, creating organic lead generation within engaged communities.

Integrate lead generation across Facebook's various features including Page calls-to-action, Group pinned posts, and Live session promotions that direct interested prospects to your email capture systems.

Nurturing Prospects Through Value Delivery

Develop email sequences that continue the relationship building started on Facebook while gradually introducing your services and demonstrating your ability to help prospects achieve their goals.

Use Facebook's remarketing capabilities to stay visible to website visitors and email subscribers, creating multiple touchpoints that build familiarity and trust over extended periods.

Create exclusive content for Facebook communities that demonstrates additional value while building anticipation for your paid offerings and services.

Share client success stories and case studies through Facebook content that provides social proof while helping prospects understand the transformation possible through your services.

Converting Community Members to Clients

Identify the most engaged community members who consistently participate in discussions, share your content, and demonstrate genuine interest in your expertise as potential clients or referral sources.

Offer exclusive opportunities to community members such as discounted services, early access to new offerings, or complimentary consultations that reward community participation while generating business opportunities.

Use Facebook Messenger for direct communication with interested prospects, providing personalized attention while maintaining the informal, social context that makes business conversations feel natural.

Track which Facebook activities generate the most business inquiries and client conversions, focusing your effort on the community building and content strategies that produce the best business results.

Systematic Business Development Process

Create documented processes for moving Facebook connections through your business development funnel, from initial community engagement through client conversion and ongoing relationship maintenance.

Develop content calendars that balance community value with business objectives, ensuring consistent service to your community while systematically promoting your expertise and availability for professional services.

Use Facebook's analytics and advertising insights to understand which audience segments and content types generate the best business results, optimizing your funnel for maximum efficiency and conversion.

Establish systems for following up with prospects generated through Facebook, ensuring consistent communication that builds relationships while identifying qualified clients ready for your services.

Action Steps for Chapter 8:

Action Item	Task Description	Expected Outcome
Facebook Group Creation and Management	Create a focused Facebook Group around your expertise area with clear guidelines, valuable content themes, and regular engagement activities. Plan group growth strategy and community building activities.	Active professional community that positions you as a thought leader while providing networking and business development opportunities
Facebook Page Optimization	Set up or optimize your Facebook Page with comprehensive business information, professional branding, clear service descriptions, and regular valuable content that builds credibility with potential clients.	Professional Facebook presence that supports business credibility and provides easy contact methods for interested prospects
Live Session Strategy	Plan and execute weekly Facebook Live sessions including educational content, Q&A formats, and behind-the-scenes insights. Develop recurring themes that build audience expectations and expertise demonstration.	Regular live interaction that builds personal relationships and demonstrates expertise in real-time, natural contexts
Lead Generation System	Set up Facebook lead generation including compelling lead magnets, email capture systems, remarketing campaigns, and nurture sequences that move prospects toward client relationships.	Systematic approach to converting Facebook engagement into business leads and client relationships
Business Funnel Integration	Create a comprehensive business development process that leverages Facebook Groups, Page content, Live sessions, and advertising to systematically generate and nurture business opportunities.	Complete Facebook-based business development system that consistently generates qualified prospects and client conversions

NINE

TikTok for Fast-Track Growth

TikTok's algorithm represents the most democratic content distribution system among major social platforms, giving every creator, regardless of follower count, the opportunity to reach massive audiences through the "For You" page. Understanding how this algorithm operates becomes crucial for professionals seeking to build personal brands and establish expertise in an environment traditionally dominated by entertainment content.

The platform's algorithm prioritizes engagement velocity, content completion rates, and user interaction signals over follower counts or account age. This means that a well-crafted professional video from a new account can outperform content from established creators if it resonates strongly with its initial audience test group.

TikTok's algorithm operates on a feedback loop system that continuously learns from user behavior to serve increasingly relevant content. For professional creators, this means that consistent, valuable content aligned with your target audience's interests will eventually find and build the right community, even in TikTok's entertainment-heavy environment.

Understanding the For You Page Distribution

TikTok's For You page functions as the primary content discovery mechanism, exposing users to creators and topics beyond their direct followers. The algorithm tests every video with small audience segments before deciding whether to distribute content to larger audiences based on early engagement signals.

Initial performance metrics within the first few hours of posting determine whether the algorithm continues promoting your content. High completion rates, immediate comments, quick shares, and profile visits signal to the algorithm that your content provides value worthy of broader distribution.

The algorithm considers viewer behavior patterns, including how long someone watches your video, whether they watch it multiple times, whether they visit your profile after viewing, and whether they engage with your other content. These behavioral signals carry more weight than simple likes or follows.

Geographic and demographic matching plays a role in content distribution, with the algorithm initially showing your content to users with similar characteristics to your existing engaged audience. This creates opportunities for targeted professional content to reach relevant business audiences.

Engagement Signals That Drive Reach

Video completion rate serves as the strongest algorithmic signal, with videos watched to completion receiving significantly more distribution than content that loses viewers midway. Professional content must hook attention immediately and maintain interest throughout to succeed on the platform.

Comments generate more algorithmic value than likes because they require greater engagement investment from viewers. Creating content that naturally prompts questions, discussion, or response increases your content's distribution potential.

Shares and saves signal high-value content that users want to revisit or recommend to others. Professional content that provides actionable insights, valuable resources, or save-worthy tips often performs well due to these engagement behaviors.

Profile visits following video views indicate strong interest in your expertise and content, signaling to the algorithm that your video successfully introduced viewers to valuable creators they want to follow and engage with further.

Content Categories and Professional Positioning

TikTok's algorithm categorizes content into interest areas, making it essential for professional creators to establish clear content themes that help the algorithm understand your niche and target audience. Consistent posting within specific professional topics builds algorithmic recognition.

Educational content performs increasingly well on TikTok as the platform's audience matures and seeks valuable information alongside entertainment. Professional expertise shared through engaging formats can achieve significant reach while building genuine authority.

Business and career advice content has found strong audiences on TikTok, particularly among younger professionals seeking mentorship and career guidance. This creates opportunities for established professionals to share expertise while building personal brands.

Industry-specific content allows for niche audience building, with the algorithm becoming more effective at finding relevant viewers as it learns from engagement patterns within specific professional communities.

CREATING SHORT-FORM CONTENT WITH HIGH VIRALITY POTENTIAL

Professional content on TikTok succeeds when it adapts business expertise to the platform's fast-paced, engaging format while maintaining credibility and providing genuine value. The challenge lies in condensing complex professional insights into compelling 15-60 second videos that educate while entertaining.

Viral professional content typically combines trending formats with valuable expertise, using popular sounds, effects, or video structures to deliver business insights in ways that feel native to TikTok's culture while standing out for their practical value.

The most successful professional TikTok creators understand that virality comes from emotional resonance combined with practical value, creating content that people feel compelled to share because it either solved a problem or provided insights they want to pass along to their networks.

High-Performance Content Formats

"Day in the life" content that showcases professional routines, work environments, or business operations provides behind-the-scenes insights that build authenticity while demonstrating expertise in action. These videos humanize professional brands while providing valuable insights.

Quick tip or hack videos that solve specific problems in under 60 seconds perform well because they provide immediate, actionable value that viewers can implement right away. Focus on single concepts rather than comprehensive tutorials.

Reaction and commentary videos that provide professional perspectives on industry news, trends, or popular content allow for timely relevance while demonstrating expertise and unique viewpoints that add value to existing conversations.

Transformation or before-and-after content that shows results, improvements, or changes creates compelling visual narratives while demonstrating the effectiveness of your methods or approaches.

Viral Mechanics for Professional Content

Hook optimization becomes crucial in TikTok's fast-scroll environment, with the first 3 seconds determining whether viewers continue watching or move to the next video. Lead with your most compelling insight, surprising fact, or bold statement.

Pattern interrupts and visual variety maintain attention throughout short videos, using quick cuts, text overlays, props, or location changes to create dynamic content that holds viewer interest better than static talking-head formats.

Trending audio adaptation allows professional content to benefit from existing viral momentum while maintaining educational value. Use popular sounds as background for business tips or adapt trending audio to professional contexts.

Cliffhanger endings and series creation encourage viewers to follow your account for additional content, building ongoing audience relationships while allowing comprehensive coverage of complex topics across multiple videos.

Content Planning for Consistency and Growth

Batch content creation allows for consistent posting while maximizing time efficiency. Plan multiple videos around single themes, locations, or setups to create several pieces of content during focused creation sessions.

Trend monitoring and adaptation help professional creators participate in viral moments while maintaining brand relevance. Set up alerts for trending hashtags, sounds, and formats that could work with your area of expertise.

Content calendars that balance evergreen professional tips with timely industry commentary create consistent value while allowing for responsive content that capitalizes on current events and trending topics.

Performance analysis and iteration based on TikTok analytics helps identify which content types, posting times, and topics generate the best engagement with your specific audience, allowing for strategic optimization over time.

TURNING TRENDS INTO AUTHORITY-DRIVEN CONTENT

The key to professional success on TikTok lies in adapting viral trends and popular formats to showcase expertise while maintaining the entertainment value that drives platform engagement. This requires understanding how to participate in TikTok culture while positioning yourself as a credible authority in your field.

Successful trend adaptation involves identifying the core appeal of popular content formats and reimagining them through your professional lens, creating content that feels familiar to TikTok users while providing unique value through your expertise and perspective.

The most effective approach balances trend participation with original thought leadership, using trending formats as vehicles for sharing valuable professional insights rather than simply copying popular content without adding substantive value.

Trend Analysis and Professional Application

Monitor trending hashtags, sounds, and video formats to identify opportunities for professional adaptation that feel natural rather than forced. Not every trend will work with professional content, so selective participation maintains authenticity.

Analyze why certain trends become popular, whether through emotional appeal, practical value, entertainment factor, or shareability, then identify how these success factors can be applied to your professional content.

Time trend participation strategically to maximize reach while trends are still gaining momentum rather than participating after interest peaks. Early adoption often generates better results than late-cycle participation.

Maintain your professional brand voice and values while participating in trends, ensuring that trend adaptation enhances rather than conflicts with your desired professional positioning and expertise demonstration.

Content Adaptation Strategies

Educational trend adaptation transforms popular formats into learning opportunities by replacing entertainment elements with professional insights while maintaining the engaging structure that made the original trend successful.

Behind-the-scenes trend participation shows your professional life, work environment, or business operations using popular formats, providing authenticity while demonstrating expertise in natural contexts.

Industry-specific trend adaptation takes general concepts and applies them specifically to your professional field, creating niche content that serves your target audience while participating in broader TikTok conversations.

Contrarian or unique perspective trends that challenge popular opinions or provide alternative viewpoints can generate significant engagement while establishing thought leadership, provided the perspectives are well-reasoned and valuable.

Building Authority Through Strategic Trend Participation

Consistent value delivery through trend participation builds recognition as someone who provides useful professional insights regardless of format, establishing authority that transcends individual viral moments.

Expert commentary on trending topics related to your field positions you as someone who stays current with industry developments while having valuable perspectives to share about ongoing conversations.

Original trend creation based on your professional expertise can generate significant reach while establishing you as someone who contributes new ideas to TikTok culture rather than just adapting existing content.

Cross-platform promotion of successful TikTok content helps build authority across multiple channels while maximizing the reach and impact of your trend participation and professional insights.

Long-term Authority Building Strategy

Document your expertise systematically through trend participation, creating an archive of professional insights delivered through engaging formats that demonstrate both industry knowledge and communication skills.

Build recurring content themes that become associated with your professional brand, creating expectations among followers while establishing yourself as the go-to source for specific types of professional guidance.

Engage authentically with other professionals on the platform to build community relationships while expanding your network and creating opportunities for collaboration and cross-promotion.

Track which trend adaptations generate the most business inquiries, profile visits, and follower growth from your target professional audience, focusing efforts on trend types that best serve your business development goals.

Action Steps for Chapter 9:

Action Item	Task Description	Expected Outcome
TikTok Profile Optimization	Set up a professional TikTok profile with a clear bio describing your expertise, professional headshot, and links to other platforms. Develop content themes that balance professional value with platform culture.	Professional TikTok presence optimized for both algorithm success and business credibility building
Content Creation System	Develop a batch content creation process for producing 20–30 TikTok videos efficiently. Create templates for different content types including quick tips, behind-the-scenes, and trend adaptations.	Scalable system for consistent TikTok posting that maintains quality while maximizing time efficiency
Trend Monitoring and Adaptation	Set up systems for monitoring trending hashtags, sounds, and formats relevant to your industry. Create a process for quickly adapting trends to your professional expertise while maintaining authenticity.	Strategic approach to trend participation that builds authority while leveraging viral momentum for increased reach
Engagement and Community Building	Develop a daily engagement routine including responding to comments, engaging with other professionals' content, and participating in industry conversations. Focus on building genuine relationships within your professional niche.	Active TikTok community engagement that builds professional relationships and increases content visibility
Performance Tracking and Business Integration	Monitor TikTok analytics to identify best-performing content types and optimal posting strategies. Track how TikTok engagement translates to business opportunities and adjust strategy accordingly.	Data-driven TikTok strategy that maximizes both platform growth and business development outcomes

TEN

Content Strategy Across Platforms

Creating unique content for every social media platform quickly becomes overwhelming and unsustainable for busy professionals building personal brands. The most successful creators develop systematic approaches to content repurposing that maximize the value extracted from each piece of original content while adapting the format and messaging to meet the unique requirements and audience expectations of different platforms.

Effective content repurposing goes beyond simply posting identical content across multiple platforms. It requires understanding how the same core message can be adapted, reformatted, and recontextualized to provide optimal value and engagement on each platform while maintaining consistent brand messaging and expertise demonstration.

Strategic repurposing allows professionals to maintain an active presence across multiple platforms without proportionally increasing content creation time, creating efficiency that supports sustainable long-term personal brand building and business development.

The Content Multiplication Framework

Start with comprehensive, high-value content pieces that can be broken down into multiple smaller formats. A detailed case study can become a LinkedIn article, an Instagram carousel, a YouTube video, a Twitter thread, and several individual social media posts, each adapted for platform-specific audience preferences.

Identify the core insights, key points, and valuable takeaways from each piece of original content that can be extracted and reformatted for different consumption preferences and attention spans. This approach ensures each platform receives genuine value rather than diluted versions of comprehensive content.

Create content hierarchies that move from comprehensive formats (like YouTube videos or blog posts) down to quick highlights and key insights suitable for platforms with shorter attention spans. This top-down approach maintains quality while creating multiple touchpoints with your audience.

Develop template systems for common content adaptations that streamline the repurposing process. Templates for turning long-form content into social media carousels, video highlights, or quote graphics create efficiency while maintaining professional presentation standards.

Platform-Specific Adaptation Strategies

LinkedIn repurposing focuses on professional insights, industry analysis, and business applications of your content. Take personal anecdotes from other platforms and reframe them with emphasis on professional lessons learned and business implications.

Instagram adaptation emphasizes visual appeal and lifestyle integration, taking business concepts and presenting them through visually engaging formats that show personality alongside expertise. Behind-the-scenes content works particularly well for Instagram adaptation.

Twitter/X repurposing involves distilling complex ideas into concise, impactful statements that can spark discussion. Break comprehensive content into key insights, surprising statistics, or thought-provoking questions that encourage engagement.

TikTok adaptation requires transforming educational content into engaging, quick-consumption formats that provide immediate value. Focus on single concepts from longer content pieces and present them through trending formats or engaging visual approaches.

YouTube repurposing can expand brief social media content into comprehensive educational resources, using social media posts as research validation for topics that deserve deeper exploration and detailed explanation.

Facebook adaptation works well for community-building content that encourages discussion and relationship building around professional topics, taking individual insights and positioning them as conversation starters for your professional community.

Content Repurposing Systems and Workflows:

1. **Master Content Creation**: Develop comprehensive, high-value content pieces (blog posts, detailed case studies, in-depth tutorials) that serve as source material for multiple platform adaptations. This approach guarantees that all repurposed content is based on substantial value.
2. **Platform Priority Mapping**: Identify your primary and secondary platforms based on business goals and audience preferences, then create repurposing workflows that prioritize your most important platforms while ensuring quality across all channels.
3. **Format Template Library**: Create standardized templates for common content types across platforms including quote graphics, carousel layouts, video scripts, and text post structures that maintain brand consistency while enabling efficient adaptation.

4. **Content Calendar Integration**: Plan repurposing as part of your content calendar, scheduling platform-specific versions of core content to maximize reach while avoiding audience overlap that might see identical content across multiple platforms.
5. **Quality Control Systems**: Establish review processes that ensure repurposed content maintains quality standards and provides genuine value rather than appearing as lazy cross-posting that doesn't serve audience needs effectively.
6. **Performance Tracking**: Monitor which types of repurposed content perform best on each platform to optimize your adaptation strategies and focus efforts on the most effective approaches for your specific audience and business goals.

SCHEDULING, CONSISTENCY, AND AUTOMATION

Consistent content publishing across multiple platforms requires systematic approaches that maintain regular audience engagement without overwhelming your daily schedule or compromising content quality. The most successful personal brands develop automated systems that handle routine posting while preserving authentic engagement and timely responses to audience interaction.

Effective scheduling and automation strikes the balance between efficiency and authenticity, using technology to handle mechanical aspects of content distribution while maintaining the human touch that builds genuine relationships and trust with your professional audience.

Strategic automation focuses on content distribution rather than content creation or community engagement, ensuring that your personal brand maintains the authentic interaction that social media audiences value while creating sustainable systems for long-term brand building.

Content Scheduling Strategy

Optimal posting times vary by platform and audience demographics, requiring research into your specific audience's online behavior patterns. LinkedIn audiences typically engage during business hours, while Instagram users might be more active during evenings and weekends.

Platform-specific scheduling considers not only timing but also content type preferences at different times. Educational content might perform better during weekday business hours, while lifestyle or behind-the-scenes content could generate better engagement during personal time.

Frequency optimization involves finding the sweet spot between staying visible and avoiding audience fatigue. Different platforms have different optimal posting frequencies, with Twitter supporting multiple daily posts while LinkedIn might be more effective with weekly comprehensive content.

Global audience considerations become important as your personal brand grows internationally. Consider time zones of your key audience segments when scheduling content to ensure maximum visibility and engagement potential.

Automation Tools and Best Practices

Content scheduling platforms like Buffer, Hootsuite, Later, or Creator Studio allow batch content preparation and automated posting across multiple platforms while maintaining platform-specific optimization for each post.

Automation rules should include platform-specific hashtag strategies, optimal posting times, and content format adaptations that ensure each scheduled post feels native to its platform rather than obviously automated.

Engagement monitoring systems alert you to comments, messages, and mentions that require personal response, ensuring that automation doesn't create barriers to genuine community interaction and relationship building.

Content queue management involves maintaining scheduled content pipelines that allow for timely responses to current events, industry developments, or trending topics while ensuring consistent posting during busy periods.

Maintaining Authenticity Within Automated Systems

Personal engagement windows should be built into your schedule to respond to comments, participate in conversations, and engage with your community's content in ways that automation cannot replicate effectively.

Real-time content creation allows for spontaneous posts that respond to current events, industry news, or personal insights that couldn't be scheduled in advance, maintaining the authentic, immediate feel that builds trust.

Community interaction protocols ensure that automated systems don't interfere with genuine relationship building by clearly separating automated posting from personal engagement and relationship development activities.

Content variation within automated systems prevents your feed from feeling robotic by including different content types, tones, and formats that showcase various aspects of your expertise and personality.

ANALYTICS: HOW TO MEASURE GROWTH AND IMPACT

Understanding performance metrics across different platforms enables data-driven decisions about content strategy, platform prioritization, and resource allocation that optimize both audience growth and business development outcomes. Each platform provides unique analytics that reveal different aspects of audience behavior and content effectiveness.

Effective analytics goes beyond vanity metrics like followers and likes to focus on engagement quality, business impact, and audience actions that align with your professional objectives and personal brand goals. The most valuable metrics indicate audience interest, trust building, and business opportunity generation.

Cross-platform analytics comparison helps identify which platforms generate the best return on investment for your specific goals, allowing strategic resource allocation that maximizes impact while maintaining sustainable content creation practices.

Key Performance Indicators by Platform

LinkedIn analytics focus on professional engagement including profile views, connection requests, post engagement rates, and click-through rates to your website or business offerings. These metrics indicate growing professional recognition and business opportunity potential.

Instagram metrics emphasize visual engagement including Story completion rates, Save rates, Share rates, and profile visits following post engagement. Instagram analytics also provide demographic information that helps understand audience composition and preferences.

YouTube analytics offer comprehensive data including watch time, audience retention, click-through rates, and subscriber conversion rates that indicate content quality and teaching effectiveness. YouTube metrics also show traffic sources that reveal how audiences discover your content.

Twitter/X analytics track engagement rates, profile visits, link clicks, and mention quality that indicate thought leadership recognition and industry influence. Twitter metrics also show hashtag performance and conversation participation effectiveness.

TikTok analytics focus on completion rates, share rates, and profile visits that indicate content virality potential and audience building effectiveness within the platform's algorithm-driven environment.

Facebook analytics provide community engagement data including group activity, post reach, and audience demographics that help understand community building effectiveness and business lead generation potential.

Business Impact Measurement

Lead generation tracking connects social media activity to email signups, consultation requests, and other business inquiries that represent tangible professional opportunities resulting from personal brand building efforts.

Client acquisition attribution helps identify which platforms and content types generate the most business opportunities, allowing strategic focus on activities that best support professional growth and revenue generation.

Professional recognition metrics including speaking invitations, media mentions, collaboration requests, and industry award nominations indicate growing authority and influence within your professional field.

Network growth quality assessment examines not just follower numbers but the professional relevance and engagement quality of your growing network, focusing on connections that provide business value and collaborative potential.

Strategic Optimization Based on Analytics

Content performance analysis identifies which topics, formats, and posting strategies generate the best engagement and business results, allowing optimization of future content creation efforts for maximum impact.

Audience development insights reveal demographic trends, engagement patterns, and growth sources that inform strategic decisions about platform focus, content themes, and business development approaches.

ROI calculation for social media efforts involves comparing time and resource investment against business outcomes, including lead generation, client acquisition, and professional opportunity development, to ensure sustainable brand-building practices.

Platform prioritization based on performance data helps allocate limited time and resources to platforms and activities that generate the best results for your specific professional objectives and business goals.

Action Steps for Chapter 10:

Action Item	Task Description	Expected Outcome
Content Repurposing System Development	Create a systematic workflow for turning one piece of comprehensive content into 6–8 platform-specific adaptations. Develop templates and processes that maintain quality while maximizing efficiency.	Streamlined content creation process that maintains active presence across multiple platforms without proportional time investment
Scheduling and Automation Setup	Implement content scheduling tools and establish optimal posting times for each platform. Create automated workflows that handle distribution while preserving authentic engagement opportunities.	Consistent content publishing across platforms with systems that support sustainable long-term brand building

Action Item	Task Description	Expected Outcome
Analytics Dashboard Creation	Set up comprehensive analytics tracking across all platforms focusing on business impact metrics rather than vanity metrics. Create monthly reporting systems that inform strategic optimization decisions.	Data-driven understanding of which platforms and content types generate the best professional and business outcomes
Cross-Platform Content Calendar	Develop an integrated content calendar that coordinates messaging across platforms while adapting format and emphasis for each platform's unique audience and culture. Plan content themes and campaigns systematically.	Strategic approach to multi-platform content that maximizes reach while maintaining consistent brand messaging and professional positioning
Performance Optimization Protocol	Establish regular review cycles for analyzing performance data and optimizing content strategy based on engagement patterns, business lead generation, and professional opportunity development across all platforms.	Continuous improvement system that ensures social media efforts generate maximum business value and professional recognition

ELEVEN

Monetizing Your Personal Brand

Personal brand monetization extends far beyond traditional employment or single-service business models to encompass multiple revenue streams that leverage your expertise, audience, and reputation in different ways. The most successful personal brands develop diversified income portfolios that reduce dependence on any single revenue source while maximizing the value extracted from brand building efforts.

Understanding various monetization options allows strategic development of income streams that align with your expertise, audience size, and business objectives while creating multiple pathways for professional and financial growth. Each income stream requires different approaches but benefits from the credibility and audience built through consistent personal brand development.

The key to sustainable personal brand monetization lies in balancing short-term revenue opportunities with long-term brand building activities that compound over time to create increasing value and income potential.

Sponsorship and Partnership Development

Brand partnerships provide revenue opportunities by connecting your audience with products, services, or companies that align with your expertise and values. Successful sponsorships feel authentic to your audience while providing genuine value rather than obvious promotional content.

Corporate partnerships can include content sponsorship, event partnerships, advisory roles, or brand ambassadorships that provide ongoing income while leveraging your professional credibility and audience influence for business value.

Product partnerships with companies whose offerings serve your audience create win-win relationships where your endorsement drives business results while your audience gains access to valuable resources and solutions.

Negotiating partnership terms requires understanding your audience demographics, engagement rates, and influence metrics while clearly communicating the value you provide to potential partners through authentic endorsement and professional credibility.

Speaking Engagement Opportunities

Professional speaking generates revenue while building authority and expanding your network through conference presentations, workshop facilitation, corporate training, and keynote addresses that showcase your expertise to new audiences.

Speaker fee structures vary based on event type, audience size, travel requirements, and your established reputation, with fees ranging from modest honorariums for local events to substantial fees for major conference keynotes.

Speaking topics should align with your established expertise while addressing current industry needs, challenges, or opportunities that conference organizers and corporate clients want their audiences to understand and address.

Building a speaking portfolio requires developing signature presentations, collecting video testimonials, creating professional speaker materials, and networking within conference and event planning communities.

Consulting and Professional Services

Consulting services leverage your expertise to help clients solve specific problems or achieve desired outcomes, typically commanding higher hourly rates than traditional employment while providing flexibility and diverse project opportunities.

Service packaging transforms consulting into scalable offerings like strategic planning intensives, implementation programs, or ongoing advisory relationships that provide predictable revenue while serving client needs comprehensively.

Client acquisition through personal brand activities creates inbound inquiry flow that reduces marketing costs while attracting clients who already understand your expertise and approach through your content and reputation.

Premium pricing becomes possible when your personal brand establishes clear differentiation, demonstrated results, and perceived value that justifies higher fees compared to less established competitors.

Digital Product Development

Online courses transform your expertise into scalable educational products that generate revenue without requiring proportional time investment for each sale, creating passive income potential that grows over time.

Digital resources like templates, frameworks, guides, or software tools provide immediate practical value while generating revenue from your systematized knowledge and proven methodologies.

Membership communities combine ongoing education, networking, and support into recurring revenue models that build long-term relationships while providing consistent income streams.

Book publishing, whether traditional or self-published, creates credibility assets that generate both direct revenue and indirect business opportunities through enhanced authority and visibility.

BUILDING CREDIBILITY TO ATTRACT PARTNERSHIPS

Partnership opportunities increase dramatically as your personal brand builds credibility through consistent value delivery, demonstrated expertise, and professional recognition within your industry. Companies and organizations prefer partnering with individuals who enhance their reputation while delivering authentic value to target audiences.

Building partnership-worthy credibility requires systematic development of trust indicators, including audience engagement, professional achievements, media recognition, and peer endorsements that validate your expertise and influence within your field.

The most valuable partnerships emerge from genuine relationship building rather than transactional networking, requiring long-term commitment to providing value and building trust with potential partners before commercial opportunities develop.

Establishing Professional Credibility

Thought leadership content that provides unique insights, challenges conventional thinking, or introduces innovative approaches positions you as someone worth partnering with for access to fresh perspectives and valuable expertise.

Media appearances, including interviews, article contributions, and expert commentary, build third-party validation while expanding your visibility beyond your immediate network and social media following.

Professional recognition through industry awards, certifications, speaking invitations, and peer acknowledgments provides external validation of your expertise and professional standing within your field.

Client results and case studies demonstrate your ability to generate real outcomes for others, providing potential partners with confidence that collaboration will produce measurable value rather than just theoretical expertise.

Building Strategic Relationships

Industry networking through conferences, professional associations, and peer groups creates opportunities to build relationships with potential partners while staying current with industry developments and collaboration opportunities.

Content collaboration with other professionals through joint articles, co-hosted events, or shared projects demonstrates your ability to work effectively with others while expanding your network and visibility.

Advisory roles and board positions provide opportunities to contribute expertise while building relationships with business leaders who might become collaboration partners or recommend you for partnership opportunities.

Referral relationship development with complementary professionals creates mutual support networks that generate business opportunities while providing partners with trusted resources for their clients and connections.

Demonstrating Partnership Value

Audience metrics and engagement data provide concrete evidence of your reach and influence that potential partners can use to justify collaboration investments and predict partnership outcomes.

Success story documentation from previous partnerships, client work, or collaborative projects provides evidence of your ability to deliver results and work effectively with others in professional relationships.

Professional reputation management through consistent quality delivery, reliable communication, and ethical business practices builds the trust foundation necessary for high-value partnership opportunities.

Unique value proposition articulation that clearly communicates what you bring to partnerships helps potential collaborators understand how working with you will enhance their objectives and provide competitive advantages.

CREATING A LONG-TERM BRAND MONETIZATION ROADMAP

Sustainable personal brand monetization requires strategic planning that balances immediate income needs with long-term wealth-building opportunities. The most successful personal brands develop systematic approaches that compound value over time rather than optimizing only for short-term revenue maximization.

A comprehensive monetization roadmap considers your expertise development, audience growth, platform expansion, and business evolution to create increasing value and income potential that builds sustainable professional and financial security.

Long-term thinking in personal brand monetization means making strategic decisions about partnerships, pricing, and opportunities that enhance rather than diminish your brand value while building assets that generate ongoing income beyond active work.

Monetization Stage Development

Early-stage monetization focuses on building credibility and establishing market positioning while generating modest income through services, small partnerships, or initial digital products that validate your value proposition.

Growth-stage opportunities include expanded service offerings, higher-value partnerships, speaking engagements, and scalable digital products that leverage your established audience and proven expertise for increased income potential.

Mature-stage monetization incorporates passive income streams, strategic investments, licensing opportunities, and advisory roles that generate significant income while requiring minimal ongoing time investment.

Legacy-stage planning considers how your personal brand and expertise can create lasting value through educational institutions, published works, or business assets that continue generating income and impact beyond your active involvement.

Revenue Stream Integration Strategy

Complementary income streams that support and enhance each other create more sustainable revenue than competing or conflicting monetization approaches. Speaking engagements that promote your consulting services or digital products that demonstrate your coaching methodology.

Seasonal planning that considers industry cycles, conference schedules, and business planning periods helps optimize timing for different monetization activities while ensuring consistent income throughout the year.

Investment in business infrastructure, including systems, team members, and capabilities that support multiple income streams simultaneously, increases efficiency while enabling growth that wouldn't be possible through individual effort alone.

Exit strategy consideration for various income streams allows flexibility to adapt to changing market conditions, personal preferences, or new opportunities while maintaining overall business stability and growth trajectory.

Long-term Value Building

Intellectual property development, including proprietary methodologies, trademarked processes, or unique frameworks, creates defensible business assets that command premium pricing while providing competitive differentiation.

Brand equity building through consistent quality, reliable delivery, and positive reputation creates intangible assets that support all monetization activities while increasing overall business value.

Network development and relationship building create ongoing opportunities for partnerships, referrals, and collaborative ventures that compound over time to generate increasing business value and income potential.

Knowledge and expertise expansion through continuous learning, skill development, and experience accumulation ensures that your monetization capabilities grow over time rather than becoming outdated or less valuable in changing markets.

Action Steps for Chapter 11:

Action Item	Task Description	Expected Outcome
Income Stream Assessment and Planning	Evaluate your current expertise and audience to identify 3–5 potential income streams. Research market demand, pricing structures, and requirements for each option. Create development timeline and resource requirements.	Strategic monetization plan that leverages your unique strengths while building multiple revenue sources

Action Item	Task Description	Expected Outcome
Credibility Building Initiative	Implement systematic credibility building including thought leadership content creation, media outreach, professional recognition pursuit, and case study documentation. Focus on third-party validation sources.	Enhanced professional credibility that attracts partnership opportunities and justifies premium pricing
Partnership Opportunity Development	Research potential brand partners, speaking opportunities, and collaboration prospects. Create outreach materials including media kit, speaker one-sheet, and partnership proposal templates.	Active pipeline of partnership opportunities that could generate significant revenue and brand enhancement
Digital Product Creation	Choose one digital product format (course, guide, template, or membership) and create a comprehensive offering including content development, pricing strategy, and launch plan.	Scalable income source that generates revenue without proportional time investment
Long-term Monetization Roadmap	Create a 3-year monetization plan that shows progression from current state to desired income levels through strategic development of multiple revenue streams. Include specific milestones and success metrics.	Clear roadmap for building sustainable, diversified income through strategic personal brand development

TWELVE

Scaling and Sustaining Your Brand

As your personal brand matures, strategic collaborations and joint ventures provide powerful opportunities to accelerate growth, expand reach, and create value that exceeds what you could achieve independently. These partnerships leverage combined expertise, audiences, and resources to create outcomes that benefit all parties while serving larger communities more effectively.

Successful collaborations require careful partner selection, clear value exchange, and structured agreements that protect everyone's interests while maximizing collective impact. The most valuable partnerships combine complementary strengths rather than competing capabilities, creating synergy that enhances rather than dilutes individual brand positioning.

Strategic collaboration becomes essential for scaling beyond individual capacity constraints while maintaining brand quality and authenticity. These partnerships allow access to new audiences, enhanced credibility, and expanded capabilities without compromising the personal brand foundation you've built.

Strategic Partnership Selection

Identify potential partners whose audience overlaps with but doesn't duplicate your target market, creating opportunities for mutual audience expansion without cannibalizing existing relationships. Look for professionals whose expertise complements rather than competes with your specialization.

Evaluate potential partners' brand reputation, content quality, and audience engagement to ensure alignment with your standards and values. Poor partner choices can damage your brand reputation even when collaboration intentions are positive.

Consider geographic, demographic, or industry expansion opportunities through partnerships with professionals who serve similar audiences in different markets, creating growth opportunities that leverage your expertise in new contexts.

Assess partnership compatibility through smaller collaborative projects before committing to major joint ventures, allowing relationship evaluation and working dynamic assessment without significant risk or investment.

Joint Venture Structure Development

Define clear roles, responsibilities, and contribution expectations for each partner to prevent confusion and conflict while ensuring equitable value exchange that benefits all parties appropriately.

Establish intellectual property agreements that protect individual brand assets while creating shared ownership of collaborative content and resources developed through joint efforts.

Develop revenue-sharing structures that reflect each partner's contribution while incentivizing continued collaboration and mutual success rather than creating competitive tension or resentment.

Create communication protocols and decision-making processes that ensure smooth collaboration while respecting individual brand autonomy and maintaining positive working relationships throughout the partnership.

Collaboration Value Maximization

Cross-promotion strategies that introduce each partner to the other's audience should feel natural and valuable rather than forced or purely promotional, providing genuine value to all audiences involved.

Content co-creation that brings together different areas of expertise results in valuable resources that neither partner could create alone, showcasing their teamwork and adding more value.

Event and program partnerships that combine resources and audiences can create larger, more impactful offerings while sharing costs and risks associated with significant business development investments.

Working with partners helps you connect with larger professional groups and discover new opportunities that go beyond just the current collaboration.

DELEGATING AND OUTSOURCING BRAND MANAGEMENT

Scaling your personal brand requires transitioning from handling all activities personally to building team capabilities that maintain brand quality while expanding capacity for growth and business development. Strategic delegation preserves your personal involvement in high-value activities while ensuring consistent brand representation across all touchpoints.

Effective delegation requires developing clear brand guidelines, training team members thoroughly, and creating quality control systems that maintain your brand standards while allowing others to handle routine tasks and operational responsibilities.

The goal of delegation isn't to remove yourself from your brand but to focus your personal involvement on activities that uniquely require your expertise, personality, and decision-making while building systems that support sustainable growth.

Team Building for Brand Support

Virtual assistants can handle scheduling, basic customer service, research, and administrative tasks that support your brand activities without requiring deep expertise or personal brand representation.

Content creators, including writers, designers, and video editors can develop materials under your guidance, maintaining brand consistency while expanding content production capacity beyond your personal time constraints.

Social media managers can handle posting schedules, routine engagement, and community management while escalating important conversations and relationship-building opportunities to your personal attention.

Business development support, including lead qualification, client onboarding, and project coordination, allows you to focus on high-value client relationships and strategic business development while ensuring smooth operational execution.

Quality Control and Brand Consistency

Brand guidelines documentation should include voice characteristics, visual standards, messaging frameworks, and response templates that enable team members to represent your brand authentically and consistently.

Review processes for all external communications ensure that delegated work meets your quality standards and accurately represents your expertise and personality before reaching your audience.

Training programs for team members should include brand education, communication standards, and escalation procedures that ensure appropriate handling of various situations while maintaining brand integrity.

Regular performance evaluation and feedback systems help team members improve their brand representation while identifying areas where additional training or process refinement might enhance consistency and quality.

Maintaining Personal Brand Authenticity

Reserve high-visibility activities, including media interviews, speaking engagements, and important client relationships, for your personal involvement to maintain authentic connection with your audience and key stakeholders.

Personal content creation for the most important brand communications ensures that your unique voice and perspective remain central to your brand while delegated activities handle supportive and operational functions.

Relationship management with key clients, partners, and industry contacts should remain your personal responsibility to maintain the authentic connections that drive business opportunities and collaborative partnerships.

Strategic decision-making about brand direction, partnership opportunities, and business development should involve your direct input to ensure alignment with your goals and values while leveraging team support for implementation.

BUILDING A BRAND THAT THRIVES BEYOND TRENDS

Sustainable personal brands are built on timeless principles and enduring value rather than trending topics or temporary cultural moments. While trend participation can provide short-term visibility, long-term brand success requires developing expertise, reputation, and relationship foundations that remain valuable regardless of changing platforms or popular topics.

Creating trend-resistant brand foundations requires focusing on fundamental human needs, universal professional challenges, and timeless success principles that maintain relevance across changing business environments and technological developments.

The most enduring personal brands combine consistent core values with adaptive communication methods, maintaining authentic expertise and helpful perspective while evolving presentation formats to serve audiences effectively across different platforms and cultural contexts.

Timeless Value Creation

Focus on solving fundamental problems that persist across technological and cultural changes rather than addressing only current trends or temporary challenges that might become irrelevant quickly.

Develop expertise in principles and frameworks that apply broadly across industries and situations rather than specializing exclusively in tools or methods that might become obsolete with technological advancement.

Build a reputation based on character, integrity, and consistent value delivery rather than controversial positions or trend-based attention that might damage long-term credibility and relationship potential.

Create educational content that teaches thinking processes and decision-making frameworks rather than just providing answers to current questions, ensuring ongoing value as circumstances change.

Platform Independence Strategy

Develop owned media, including email lists, websites, and direct communication channels that don't depend on social media platforms for audience access and relationship maintenance.

Create content that can be adapted across multiple platforms and formats rather than developing platform-specific content that becomes worthless if platforms change or decline.

Build direct relationships with your audience through personal interaction, events, and communication that create loyalty beyond platform convenience or algorithmic exposure.

Maintain presence across multiple platforms while avoiding overdependence on any single channel for business development or audience relationship building.

Legacy Brand Development

Document your expertise and methodology in comprehensive formats, including books, courses, or structured programs that provide lasting value and establish permanent authority in your field.

Mentor and develop other professionals who can extend your influence and carry forward your expertise contributions while building network relationships that support long-term business sustainability.

Create intellectual property, including frameworks, processes, or systems that can provide ongoing value and income potential while establishing lasting contributions to your professional field.

Build institutional relationships with organizations, educational institutions, or industry groups that provide stability and ongoing influence opportunities beyond individual social media presence or trending popularity.

Evolution and Adaptation Strategy

Keep an eye on changes in the industry and technology that could impact what your audience needs, while still focusing on your main skills and the value you offer that go beyond just specific tools.

Continuously develop new skills and knowledge areas that complement your established expertise while expanding your ability to serve evolving audience needs and market opportunities.

Regular brand evaluation and strategic planning help find changes that need to be made while keeping the core positioning and value delivery the same, which helps build long-term relationships.

Succession planning considerations for your personal brand, including knowledge transfer, relationship maintenance, and asset preservation, ensure that brand value extends beyond your individual involvement and contributes lasting value to your professional field.

Action Steps for Chapter 12:

Action Item	Task Description	Expected Outcome
Strategic Partnership Development	Identify 5–10 potential collaboration partners whose expertise complements yours. Research their audience alignment and brand reputation. Develop partnership proposals for joint content, events, or business ventures.	Active collaboration opportunities that expand reach and capabilities while maintaining brand integrity
Team Building and Delegation System	Define roles that can be delegated while maintaining brand quality. Hire and train virtual assistants or team members for content creation, social media management, and administrative support. Create quality control processes.	Scalable brand management system that expands capacity while preserving authenticity and quality standards

Action Item	Task Description	Expected Outcome
Brand Guidelines and Training Materials	Create comprehensive brand guidelines including voice characteristics, visual standards, response templates, and quality control processes. Develop training materials for team members who represent your brand.	Systematic approach to maintaining brand consistency as you scale operations and delegate responsibilities
Platform Independence Strategy	Develop owned media channels including email list, website, and direct communication methods. Create content systems that work across multiple platforms. Reduce dependence on any single platform for audience access.	Sustainable brand foundation that thrives regardless of platform changes or algorithm updates
Legacy Planning and Documentation	Begin documenting your expertise in permanent formats including comprehensive guides, frameworks, or educational programs. Plan for long-term brand sustainability and knowledge transfer.	Enduring brand assets that provide lasting value and establish permanent authority in your professional field

Your Words Could Inspire the Next Great Personal Brand

At the heart of every act of generosity is a simple truth: when we share what we know, we help others grow.

Have you ever thought about how even a small piece of encouragement or guidance can change the course of someone else's life?

Picture a reader standing at the starting line—eager to build a personal brand but overwhelmed by the maze of **LinkedIn, X, Instagram, YouTube, Facebook, and TikTok**. They may have valuable skills or stories to share but lack the confidence or clarity to put themselves out there. Your reflections on *Personal Branding: A Complete System to Build and Monetize a Unique Brand Using Influencer Strategies and Step-by-Step Marketing Tools on LinkedIn, X, Instagram, YouTube, Facebook, and TikTok* could be the nudge that convinces them it's possible.

That's why I'd be so grateful if you would leave an **honest review**. It's more than feedback—it's part of a shared mission to help others discover their voice, grow their influence, and take bold steps toward turning passion into opportunity.

By sharing your thoughts, you:

- **Guide future readers** who wonder if this book can help them rise above the noise.
- **Support the vision** of helping more people build brands that feel authentic and meaningful.
- **Amplify impact**, because your words can reach far beyond a single review.

Thank you sincerely for taking the time to share your perspective. And if you know someone who dreams of creating a brand that opens doors and inspires others, I'd be honored if you passed this book along to them as well.

Conclusion

The rules of business have fundamentally changed. Gone are the days when hiding behind corporate logos and company brands was enough to build a successful career or business. Today, people buy from people, not faceless corporations. They follow individuals who share their values, solve their problems, and inspire their aspirations.

THE SHIFT FROM CORPORATE IDENTITY TO PERSONAL IDENTITY IN BUSINESS

Traditional business operated on institutional trust. Companies spent decades building brand recognition through advertising, public relations, and corporate messaging. Employees were expected to be interchangeable parts of the corporate machine, their individual personalities absorbed into company culture and brand guidelines.

This model has become obsolete in the digital age. Consumers now expect transparency, authenticity, and personal connection from the brands they support. They want to know who stands behind the products and services they purchase. They seek relationships, not just transactions.

Consider how business decisions are made today. When choosing between two consultants with similar credentials, clients almost always select the one whose personal brand resonates with their values and communication style. When investors evaluate startups, they are betting on the founders' personal brands as much as the business concept. When companies hire executives, they consider not only professional qualifications but also how that person's personal brand aligns with and enhances their organizational goals.

The most successful professionals understand that their personal brand is their competitive advantage. It is what differentiates them in crowded markets, attracts opportunities, and commands premium pricing. Your personal brand becomes your business moat, something competitors cannot replicate because it is uniquely yours.

WHY PERSONAL BRANDING IS ESSENTIAL IN THE DIGITAL AGE

Digital platforms have democratized influence and entrepreneurship in ways never seen before. Anyone with expertise, passion, and strategic thinking can build a global audience and generate significant income. However, this same accessibility has created intense competition for attention and trust.

Without a strong personal brand, you become invisible in the digital noise. Your expertise remains hidden, your potential clients cannot find you, and opportunities pass to those who have mastered the art of personal branding. The choice is not whether to build a personal brand, but whether to build one intentionally or allow others to define it for you.

Search engines and social media algorithms favor individuals who consistently share valuable content and engage authentically with their communities. They reward personal brands that demonstrate expertise, build relationships, and create genuine value for their audiences. These platforms have become merit-based systems where the most helpful, authentic, and consistent personal brands rise to the top.

Modern consumers conduct extensive research before making purchasing decisions. They examine your online presence, read your content, watch your videos, and evaluate your credibility before engaging with your services or products. Your personal brand serves as your twenty-four-hour sales representative, building trust and demonstrating value even when you are not directly involved in the conversation.

The global nature of digital business means your personal brand can attract opportunities from anywhere in the world. Geographic limitations no longer constrain professional growth when your expertise and personality can reach international audiences through strategic personal branding.

WHAT THIS BOOK PROVIDES: A COMPLETE SYSTEM FROM BUILDING TO MONETIZING

This book provides a comprehensive, step-by-step system for building and monetizing a personal brand across all major digital platforms. You will learn not just what to do, but how to do it efficiently and effectively, with specific strategies tailored to each platform's unique characteristics and audience expectations.

The system covers every aspect of personal branding: discovering your unique value proposition, crafting compelling messages, designing professional visual elements, building authority across multiple platforms, creating engaging content consistently, and converting your influence into sustainable income streams.

Each chapter builds upon the previous one, creating a cohesive strategy that compounds your efforts rather than scattering them across disconnected tactics. You will understand how content created for one platform can be adapted and repurposed for others, how different platforms serve different functions in your overall brand strategy, and how to create synergies that amplify your impact.

The strategies presented here are based on proven methods used by successful personal brands across various industries. They are practical, implementable, and designed to generate real results rather than just theoretical understanding. Every concept is accompanied by specific action steps that allow immediate implementation.

READER PROMISE: TRANSFORM VISIBILITY INTO INFLUENCE AND INFLUENCE INTO INCOME

By the end of this book, you will have a complete roadmap for transforming from hidden expert to visible authority to profitable influencer. This transformation does not happen overnight, but it follows predictable patterns that you can accelerate through strategic action and consistent execution.

You will understand how to identify and communicate your unique value in ways that resonate with your target audience. You will know how to create content that builds trust, demonstrates expertise, and naturally leads to business opportunities. You will have specific strategies for each major social platform and understand how to integrate them into a cohesive personal branding system.

Most importantly, you will have a clear path for monetizing your personal brand through multiple income streams including consulting, speaking, digital products, sponsorships, and partnership opportunities. You will understand how to price your expertise, attract premium clients, and build a sustainable business around your personal brand.

The journey from anonymity to influence requires commitment and consistent action, but the rewards extend far beyond financial success. A strong personal brand provides professional security, creative fulfillment, and the satisfaction of helping others while building something uniquely yours.

Your expertise has value. Your perspective matters. Your audience is waiting for someone exactly like you to help them achieve their goals. This book provides the systematic approach to connect your unique value with the people who need it most, creating mutual benefit that forms the foundation of sustainable business success.

> *"The best time to plant a tree was 20 years ago. The second best time is now."*
>
> – Chinese Proverb

Your personal brand empire starts today.

PERS

www.ingramcontent.com/pod-product-compliance
Lightning Source LLC
Chambersburg PA
CBHW052354280926
52269CB00030B/490
* 9 7 8 1 9 7 1 0 7 1 9 0 9 *